PRAISE FOR *ARRIVING TODAY*

"Our global economy runs on logistics. Mims expertly demystifies this secretive science as he vividly portrays the ways in which it often robs the most vulnerable workers of their health and humanity."

—Brad Stone, author of *The Everything Store and Amazon Unbound*

"With the elegance and efficiency of a first-rate tech journalist, Mims leads us into the nooks and crannies, robots, AI, warehouses, and ships that are highly complex so as to make our daily life simple. A must-read."

—Scott Galloway, professor of marketing at NYU Stern School of Business and author of *The Four* and *The Algebra of Happiness*

"A meticulously and presciently rendered account of the surprising journey of a USB charger from the factory to my home. It's nice to get your stuff fast. But Mims asks us to ponder, Was it worth it?"

—Steve LeVine, author of *The Powerhouse*

"Adeptly draws us into the container ships, fulfillment centers, and algorithms that deliver us what we want, when we want it. A balanced, much-needed account."

—Robert Kanigel, author of *The One Best Way* and *Hearing Homer's Song*

"Mims elegantly explores the micro and the macro of how our modern world of stuff works, in a way that illuminates, dazzles, and sometimes terrifies."

—Rose George, author of *Ninety Percent of Everything, Nine Pints,* and *The Big Necessity*

"A backstage pass into the twenty-first-century global economy, *Arriving Today* is the resource for understanding how modern supply chains really work—and why they sometimes fail."

—Ryan Petersen, CEO of Flexport

"Finally, a book that sheds light on automation, logistics, and their impact on our everyday life today, and in the future. An engaging and insightful narrative."

—Oren Etzioni, professor emeritus at the University of Washington and CEO of the Allen Institute for Artificial Intelligence

"*Arriving Today* is the essential key to understanding how our world is getting smaller and more interconnected by the day."

—Garry Tan, cofounder of Initialized Capital

"Captures the reality, complexity, and human side of the final mile— generally the largest cost for logistics companies—and the reader gets a real glimpse into what it takes."

—Jack Levis, senior director of industrial engineering, UPS

ARRIVING

TODAY

FROM FACTORY

TO FRONT DOOR—

WHY EVERYTHING

HAS CHANGED

ABOUT HOW AND

WHAT WE BUY

ARRIVING TODAY

CHRISTOPHER MIMS

HARPER BUSINESS

An Imprint of HarperCollins*Publishers*

HarperCollins books may be purchased for educational, business, or sales promotional use. For information, please email the Special Markets Department at SPsales@harpercollins.com.

FIRST EDITION

Designed by Elina Cohen

Library of Congress Cataloging-in-Publication Data

Names: Mims, Christopher, author.
Title: Arriving today : from factory to front door—why everything has changed about how and what we buy / Christopher Mims.
Description: First edition. | New York, NY : Harper Business, an imprint of Harper Collins Publishers, [2021] | Includes bibliographical references and index.
Identifiers: LCCN 2021018564 (print) | LCCN 2021018565 (ebook) | ISBN 9780062987952 (hardcover) | ISBN 9780062987969 (ebook)
Subjects: LCSH: Delivery of goods, | Shipment of goods. | Business logistics. | Electronic commerce. | Amazon.com (Firm)
Classification: LCC HF5761 .M56 2021 (print) | LCC HF5761 (ebook) | DDC 388/.044—dc23
LC record available at https://lccn.loc.gov/2021018564
LC ebook record available at https://lccn.loc.gov/2021018565

21 22 23 24 25 LSC 10 9 8 7 6 5 4 3 2 1

For Shep

CONTENTS

PROLOGUE

If I've done my job, the book you're holding is a skeleton key. It should help you unlock insights into the past, present, and future of automation and work. It should provide a window into the lives of the people who are most affected by the rapidly changing ways we buy, sell, and transport the billions of dollars' worth of stuff we consume every day. And it should tell you something about your own life—why everyone remains so busy despite the conveniences and instant gratification we have created for ourselves.

These might seem like unusual topics for a book that traces the path of an everyday object from the factory in which it was created, in Southeast Asia, to the front door of a home in the United States. But it turns out that something as seemingly mundane as next-day delivery could not happen without many of the most transformative technologies developed in the past one hundred years. Explaining how tapping a button on your phone yields pretty much any consumer good you could want, at your doorstep, within twenty-four hours also necessitates explaining how all the innovations that make it possible—and the people who use them—come together in a planetary-scale clockwork mechanism whose behavior is impossible to understand without building it up from its smallest constituent parts.

The landscape of technologies and systems we'll traverse ranges from the sublime—the solution of impossible mathematical problems, the nanotechnology of microchips—to the mundane—the perfection of highways, shipping containers, the architecture of large ships, management of ports, and the speedup of all manual and mental labor.

Along the way, you'll become convinced, I hope, of this astonishing fact: You live inside a factory. We all do. And you are also a worker inside that factory. As are we all. When any one of us orders something online and has it delivered to our door, we are making ourselves the end point of a conveyance system pioneered by, among others, Henry Ford. Just as he optimized the flow of parts and automobiles through his factories, we have all optimized the flow of goods, both necessary and aspirational, into our homes and lives.

In the twenty-first century, how things get to us matters as much as how they're made. With the manufacturing of even a single object spread across ever more intermediary stages, factories, and countries, in many ways the supply chain and the factory floor are now indistinguishable. Adding you, the consumer, to the equation and molding your behavior to make it more compatible with this system, through algorithms and marketing tricks, is trivial compared to all the effort that comes before you click the Buy button.

By dumb luck, I managed to put myself at the start of the journey of an item, in a factory in Vietnam, at the very moment the global Covid-19 pandemic was just getting started in China. The pandemic stretched the supply chain described in this book to its very limit and beyond. Trends that might have taken a decade were compressed into a span of a few months. Virtually overnight, we were all checking our phones daily, looking for that little notification: Arriving Today. So, to add yet another thread to this book, you also get to see the accelerated transformation of the world's supply chains from the inside, as it happened, and as I witnessed it firsthand.

Where before the odd item here or there might have been delivered to our doors, with the onset of lockdowns, suddenly it was almost everything. From toilet paper to sweatpants, webcams to groceries, whatever we could order without going outside, we did. Simultaneously, millions of people who lost their jobs switched careers and became part of the supply chains of the next-day delivery upon which the rest of us relied. This book is the story of both sides of that equation.

It explores the lives of people who make supply chains possible and of those who designed them in the first place. As such, it's also an exposition of the underlying drivers of America's growing inequality in wealth, income, and rights in the workplace.

For those of you who have enjoyed my weekly column on technology in the *Wall Street Journal*, there is also in the pages that follow a surplus of technology, described with an attention to detail that will thrill some and have others skipping ahead. (Please do: This is a long book, and it's my hope that some of its sections will be worth the price of entry for those of you with narrower interests than the entire saga it chronicles.)

You will learn what technology your cell phone has in common with spacecraft, cruise missiles, and the navigation systems of Polynesian explorers. You'll also find a (hopefully) accessible explanation of the "thinking" process of the AI (artificial intelligence) that drives an autonomous vehicle. You'll learn why automated warehouses are like microchips that process stuff instead of bits, and how the two were designed with the same principles in mind.

You will be introduced to "Bezosism," that braiding together of management practices, AI, workplace surveillance, robots, and hard automation that is the engine of Amazon's success, and possibly the future of all low-skilled labor. You will come to understand how its antecedent, "Taylorism," disappeared from our history books because it became the water in which we swim—the dominant ideology of the modern world and the root of all attempts at increasing productivity, both at work and at home.

You will meet a cast of sympathetic and not-so-sympathetic characters, from a long-haul truck driver who cannot escape his way of life no matter how good he is at his job to the architects of the systems that keep workers like him from bettering their lot. You'll meet associates in Amazon warehouses who can help answer the question, If Amazon is such a tough place to work, how has it managed to expand its ranks to more than a million full-time employees? You'll cross the Pacific with a ship's mate, take a death-defying leap from a pilot boat onto the slippery ladder hanging from the side of a container ship as big as a skyscraper, and glimpse the world's largest robot.

You will travel through time, from the often obscure roots of the systems that make our modern logistics networks possible, through a sprawling and chaotic present, and into a future in which you'll meet self-driving trucks, autonomous delivery bots, and the "intelligent" robotic arm that may someday be looked on as the progenitor of the age of ubiquitous robot laborers we've long been promised.

By the time you finish this book, I hope that you will never again be able to take a package from your front step without feeling a tiny shiver at the gobsmacking effort and complexity behind its delivery to your home.

There is no longer any such thing as a mundane journey for goods, only everyday miracles we fail to appreciate because their inner workings are hidden from view. Understanding logistics, a field too long ignored by otherwise well-informed readers, and how it has become perhaps the most sophisticated field of human endeavor, I hope you can begin, as I have, to comprehend how our modern world works, and how it might continue to evolve.

Chapter 1

THE GATHERING STORM

On an equatorially sunny day in southern Vietnam, our banana-yellow speedboat carves a broad white wake into the Saigon River. Five of us—one pilot, one sailor, two port executives, and me, ever the hapless, bespectacled journalist—shout at one another over the roar of the engine and the wind. We race a lazy current carrying lilies and the occasional glob of trash, tracing the river's sinuous curves, headed for its terminus in the South China Sea.

Jan Bandstra, a vigorous, middle-aged Dutchman who speaks "six or seven" languages and has managed ports in as many countries, observes that this must have been the same view American GIs had as they patrolled in PT boats, exchanging machine-gun fire with Vietcong hidden in the dense jungle that still lines both shores.

Soon after leaving the dock, we encounter barges with broad, V-shaped prows transporting shipping containers between small inland ports and their larger cousin abutting the ocean. We slalom between them, something nimble and out of place in this expanse of large things moving implacably.

The Leviathan that is the global economy has many hearts, each one a complex of ports, rail hubs, and highways pushing almost everything we consume across oceans and continents. We are in the antechamber of one of them. I am here for an audience with giants—cranes as big as buildings, ships as big as skyscrapers, conurbations of stacked shipping containers like cities unto themselves.

Meanwhile, a storm is brewing.

The day before, on January 20, 2020, the first case of coronavirus was confirmed in the United States. Three days after that, Wuhan, population 11 million, the capital of Hubei Province and the most important city in central China, will be locked down completely as the Chinese government races to enact the most draconian quarantine in history, including a complete prohibition of movement of anything but critical goods and persons. Eventually, it will expand to include all of Hubei Province and other areas, encompassing more than 60 million people.

Despite China's warnings, not to mention decades of research by public health specialists and intelligence analysts all over the world, by March 2020 the failure to prepare for the coming pandemic on the part of nearly all governments outside of East Asia will cause global infection rates to skyrocket—in some countries, again and again. An explosion of illness in Italy and a near collapse of its health system, anecdotes about even the young and healthy being cut down in the prime of their lives, the knowledge that there may be no cure in time to save anyone, lead to another sort of pandemic, one of fear.

Gripped by a universal anxiety of a sort unknown since the capitulation of France and the specter of a fully Nazified United States of Europe, citizens will be told that they will have to stay home, that anyone is a potential vector, that they should prepare to "shelter in place" as if what is coming is not an invisible microbe but a hurricane, or the fallout from a nuclear blast.

And, oh, add the authorities, almost as an afterthought: you should probably have enough groceries to last for at least a couple of weeks.

Confronted by the stark reality of their powerlessness to do anything else and primed by a lifetime of consumerism into thinking the answer to the existential dread at the core of their being is to buy more stuff, Americans, along with everyone else on Earth with the means to do so, will go shopping.

They will buy things that make sense, like pasta, and things that do not, like adult onesies. Millions will read and for the first time take the advice of preppers, a subculture of survivalists whose well-stocked underground bunkers, remote "bug-out" cabins, and belief in the imminent collapse of civilization bring their paranoia into vogue every time it seems the thin veneer of civilization will crack open to reveal the Hobbesian reality coursing just beneath its surface.

There will be a run on beans and eggs and milk. Hand sanitizer and disinfectant wipes will become so valuable that black markets for them arise overnight, penetrating the national consciousness as pandemic profiteers gouge the quavering masses on Amazon, only to be shut down by a strongly worded edict from its jacked and newly divorced CEO, Jeff Bezos. Later, there will be runs on other items, like desks and web cameras and laptops, as members of the professional-managerial class are forced to work from home while their children, whose heretofore overstuffed lives mean they cannot possibly be left to their own devices, learn from teachers on the other side of Google Classroom conferences and Zoom calls.

As the first week of quarantine gives way to the second and third, diapers, tampons, flour, bread yeast, and any number of other items people either need or think they do will run out. In mid-March, Amazon will announce it will stop accepting all "nonessential" items from third-party sellers in its warehouses and focus solely on acquiring and distributing the items most essential to everyday life, like food and personal-health products. Within a year, the company will have increased its frontline workforce by half, pushing it past the million-employee mark.

Amazon's vaunted promise to consumers, made only the summer before, that something like 10 million different items would be at their doorstep within one day will break down completely. The moment when an email or push alert from the company cheerfully announcing an order is "Arriving Today" begins to slip further into the future.

Delivery dates for Prime-eligible items push out to a week, then a month, then snap back to as few as two days for items deemed essential, as the colossal supply chains of the world's biggest e-commerce company stretch, convulse, and are re-formed by drastic and emergency measures.

Then things get worse.

The same supply chains that could not provide everyday consumer goods will prove equally incapable of delivering enough medical supplies to protect the frontline health-care workers who are now tasked with attending to a tidal wave of Covid-19 patients, their numbers eclipsing all efforts to test, trace, and lock down what is by now a pandemic beyond all means of conventional public health response. Some patients, it will later be revealed, are shedding virus at a rate far higher than those infected with comparable pathogens, their every breath a fountain of aerosolized infectious agents. Every one of them is both a biohazard in need of complete isolation, and all the tools and material that requires, and a human being in a state of wide-eyed, animal panic who might need sedation, intubation, and life-saving care.

In the country that invented the N95 mask, the only way to get them will be from shady third parties who bid their price to the highest possible level, in a system set up by a president who spends months downplaying the potential impact of the virus and then leaves all fifty states to battle one another for critical supplies.

In what will be compared, over and over again, with previous war-time mobilizations, American automobile manufacturers who once built 4-million-square-foot factories to churn out bombers to level the cities of the Axis powers remember their old reflexes. Partnering with health-care companies, Ford and GM will shift their factories to the construction of ventilators. Meanwhile, shoe manufacturers will pivot to making masks. Then everyone will pivot to making masks, first in America's factories and then in their own homes, from patterns downloaded from the internet and YouTube videos and tutorials in the editorial pages of the country's most read newspapers.

Still, it will not be enough to stop the virus.

As nonessential businesses remain shuttered and those that remain open begin limiting how many people can enter at any one time, long lines will form outside America's big-box meccas, turning trips to the grocery store into masked, socially distanced processions in which everyone eyes one another solemnly, knowing that nothing will ever be quite the same again.

Hospitals will run out of supplies. First to go are disposable zip-up biohazard suits, then masks. Then they will run out of ventilators, then the sedatives that keep those lucky enough to be on ventilators from panicking as tubes are inserted in their throats to keep them from perishing from a lack of oxygen, as their lungs struggle against the antibody-filled broth that fills them so completely that on X-rays they take on the characteristic translucence of ground glass.

The governor of the state of New York will go on TV every day to talk about the virus. The president of the United States will appoint his son-in-law to a shadow team intended to manage the government's response, sideline his own Centers for Disease Control and Prevention, and utter falsehoods that must immediately be corrected in official releases, as the chief of his scientific and medical response team shakes his head in the background of one of his national addresses. In states like Florida, peaks of infection will rise to heights not seen even in New York.

Food banks and state unemployment systems will experience record demand. The economic fallout will quickly appear to be even worse than the direct effects of the pandemic itself.

People will die, by the tens of thousands, and eventually by the hundreds of thousands, for lack of some critical item or another, be it the reagent to conduct a test for coronavirus, an item of personal protective equipment to stop the chain of infection three interpersonal leaps ago, or just a bed in which to convalesce.

▶ ▶ ▶

When the coronavirus hammered the global economy, it hit supply chains hardest of all. For the first time since the oil crisis of the 1970s, the world experienced a "supply shock"—a shortage of goods, rather than a dearth of demand for them. Supply shocks cannot be fixed with rate cuts by central banks or floods of capital into the economy, as with other crashes. Supply shocks are what happens when there aren't goods to be had at any price.

Those long, invisible supply chains snaking from our front doorsteps across the country, across oceans, all the way to factories 12,000 or more miles away, the ones many in the West had taken for granted almost since the day they were built, suddenly became terribly, viscerally apparent.

Almost overnight, people in the United States woke up to the fact that "made in China" didn't mean just cheap trinkets or nice-to-have luxury electronics. It also applied to things with life-and-death consequences, like N95 medical masks and the chemicals required to manufacture antibiotics and other drugs. It meant materials required for 17,600 different products made by Procter & Gamble, everything from detergent and razors to cold medicine and shampoo. It was artificial sweeteners for Diet Coke, it was the iPhone and iPad.

Even before the dust cleared and the economic carnage could be tallied—a process that will not be finished even years after this book is published—it became clear that three in four U.S. firms would have their supply chains disrupted in some way.

No one knew all of this was coming, least of all me, on January 21, 2020, the day I stepped onto a boat with Mr. Bandstra. While the node in the supply chains of Asia I was about to tour had already been growing in importance and was fast becoming something like the next China—a hub of low-cost manufacturing rapidly transforming into a hot spot of high-tech expertise—what I couldn't have predicted was that the black swan of the coronavirus was about to make everything I would see that day, and in the days following, so much more important.

Just as the Saigon River on which we traveled had shifted its course across the Mekong Delta countless times before, eradicating one shoreline as it built up another, its path dictated by forces beyond human scale, the supply chains that had until the global pandemic of

2020 seemed adequate were about to whipsaw into new configurations, a flood of goods reshaping them to fresh and urgent needs.

The day I stepped from our yellow speedboat onto the concrete pier of Cai Mep International Terminal, part of one of the largest container ports in Southeast Asia, shipping containers in its yard contained a not insignificant portion of the goods that people, businesses, and hospitals across the globe would be panic-buying in the months ahead. There were containers full of surgical masks and other personal protective equipment (PPE), and containers full of items no one had anticipated anyone would panic-buy, like Samsung refrigerators and freezers. There were also containers full of the miscellany of personal electronics required to work from home, like external monitors, power supplies, USB chargers, wireless headphones, and dongles for connecting one sort of cable to another.

It was all there, waiting on the docks, about to become part of history. And I had no idea. I thought I was there to write about an obscure but, to me, important topic: how things get from the factory, mostly in Asia, to the front doors of the homes and offices in the biggest consumer economies in the world, and specifically my own country, the United States.

But why go to Southeast Asia and not China, which is still, by volume, the number one producer of countless goods? Long before the coronavirus pandemic swept the globe, suppliers to Apple, Nike, and dozens of other multinationals had set up factories in Vietnam that are copies of their factories in China. Their reasoning, as one shipper confided to me, was simple: if tariffs or a shooting war suddenly made it impossible to get goods out of China, these companies need to be able to shift their manufacturing within days. The alternative, losing their ability to make goods at all for the months or years it would take to set up new manufacturing complexes in other countries, was a risk they felt they could not take.

Southeast Asia, and Vietnam in particular, is therefore a place that represents the future of manufacturing, as well as a much bigger portion of its present than most people realize. Many goods you have in your home and workplace are manufactured there already, from kitchen appliances and clothing to smartphones and AirPods. The shocks to the supply chain that were to come would only accelerate this dispersion of manufacturing capacity across the globe, wresting from mainland China its monopoly on the creation of countless objects.

This is not to say that the world's multinationals, and Vietnam itself, were ready. In the immediate aftermath of the black dawn of the coronavirus, manufacturers tested the hypothesis that factories outside China could protect them from shocks to global supply chains, and in many cases found it wanting. What so many companies couldn't get around, no matter where their factories were located, was that the parts and materials that are fed into today's factories still so often come from China. Manufacturing in the twenty-first century isn't material in, finished products out, as it was in the days of Bethlehem Steel and Henry Ford. Today's manufacturing is waypoints on much longer supply chains, a string of factories transforming raw materials into parts and subassemblies before final assembly in some other facility.

Perhaps the most famous example of the sprawling nature of modern manufacturing and supply chains is the fact that cod caught off Scotland are transported to China to be filleted, then transported back to Scotland for sale in grocery stores. But a 20,000-mile journey for Scottish cod is but a puddle jump compared to the cumulative supply chain miles racked up by the average smartphone, which may include lithium from Australia transformed into batteries in Korea, microchips fabricated in Taiwan from silicon grown in Japan out of quartz harvested in Appalachia, shatterproof glass manufactured in Kentucky from sand mined in Minnesota, lasers manufactured in Texas, and as many other sources, and individual supply chains, for the more than 300 components in such a device.

Every time we glance at our phones, we hold in our hands something containing more stories than could be explored in a lifetime. And a surprising number of them involve this part of the world, this still-obscure-to-outsiders corner of Southeast Asia.

Chapter 2

THE BOX

There was another reason to begin this journey in Vietnam, an un-expected historical connection between globalization and this country. The Vietnam War was possibly the most well-orchestrated and skill-fully managed bad idea in history. It killed up to 2 million Vietnamese and more than 58,000 Americans. It was also, in a strange way, the crucible in which critical technologies that have enabled our modern world, and especially modern trade, were forged.

All the online shopping we take for granted, all the globalization that seemed like a good idea at the time, much of the populism it bred—none of it would have been the same without that tragic war and the stunningly beautiful, rapidly growing, intoxicatingly vibrant place that is modern Vietnam.

Here's the connection: in its demand for materiel to supply what would eventually be a half million U.S. troops, the Vietnam War led to military contracts that subsidized and accelerated the development and adoption of the modern shipping container.

From cheap goods at Walmart and the gutting of America's mid-dle class to the rise of globalization and the elevation from poverty of more than 2 billion humans in the developing world in the past twenty years alone, so many threads of modernity run through the shipping container, and therefore Vietnam. Phenomena this gigantic have as many causes as we care to name, of course. But one of them, a big one in fact, is the demands of a conflict that was by some measures more rapacious in its consumption of gear and bombs than all of World War II.

As Marc Levinson recounts in *The Box*, his obsessively detailed history of the shipping container, "It took a major war, the United States' painful campaign in Vietnam, to prove the merit of this revolutionary approach to moving freight." Until that war, containerized shipping as we know it today, with all its interdependencies of ships, cranes, docks, trucks, railroad cars, and countless other bits of equipment, all built to accommodate standardized shipping containers, was proving nearly impossible to jump-start. Everything was a chicken-and-egg problem—huge investments in onshore cranes couldn't be justified without equally enormous investments in the proper ships. And it was compounded by a collective-action problem—the commercial interests of a variety of companies all wanting their size and type of shipping container to become the standard.

As is so often the case, investment by a government, able to generate sufficient demand all on its own, was required. Very quickly, the quagmire of "break-bulk shipping" for the war, in which gear—451 tons of it for every infantry battalion—had to be manually loaded and unloaded from ships, was resolved by shipping containers and shipping-container-accommodating ships, all provided by Sea-Land, the company founded by containerized shipping pioneer Malcolm McLean.

Once the internet arrived, containerized shipping combined with the PC, internet, and mobile revolutions to give us e-commerce and the most transformative company on the planet today, Amazon. (If Jeff Bezos makes good on his promise to build the rockets that will make orbiting space colonies possible, a generation or ten from now someone will add yet another improbability to this chain of events.)

Amazon in turn has fundamentally changed how we buy pretty much everything online and, in ways that aren't always apparent, in stores. The arrival of coronavirus accelerated its dominance and the importance of e-commerce in general, as waves of store closures and the collapse of debt-burdened retail chains, both before coronavirus and after, has evidenced.

Just as important, Amazon has perfected the kind of systems that now dictate what happens during the majority of the waking hours of most blue-collar Americans. And these same systems are beginning to take over the world of white-collar work as well.

Amazon has accomplished all this by updating old systems of management, automation, and regimentation of work. Under the leadership of Bezos and his deputies, they have become shiny, new, surveillance-powered systems of management, automation, and regimentation of work. More productive than ever, these systems turn their constituent parts, including the humans without which they could not function, into what can only be described as cyborgs.

But we are getting ahead of ourselves.

▶ ▶ ▶

It's the dry season in Vietnam, which means it's hot but it's not raining. Still, there is water everywhere, the kind that could give you typhus if you drink it. All this water is the reason Vietnam has always been so productive agriculturally. While it sometimes gets in the way, it's also one reason the country is now booming economically. From 1993 to 2016, the proportion of Vietnamese who lived in poverty dropped from 51 percent to 10 percent.

Contemporary Vietnam is like Japan in the 1960s, or China's "special economic zone" of Shenzhen, now the world's preeminent electronics manufacturing hub, in the late '90s. Cheap labor is being employed by foreign companies, many of them Chinese, Korean, or Japanese, to do the manufacturing that has become too expensive within the borders of their own countries. As rising wages in the United States inspired companies to outsource to the original cohort of "Asian tigers," so too have the economic booms of those countries raised standards of living and led workers to demand higher wages. That's pushed manufacturing down to the next and most geographically proximate countries on the economic ladder, including India, Thailand, the Philippines, and Vietnam.

All the stuff required to make finished goods has to get into Vietnam's factories somehow, and those products must also be delivered to the rest of the world in an orderly fashion. Almost anywhere else in the world, that would mean a network of highways and train tracks. Vietnam has a population of 90 million, squeezed into an area the size of New Mexico. Its borders enclose a long, narrow tendril of mountains and coastline more than a thousand miles from top to bottom, or nearly the distance from San Diego to Portland. Yet, considering its size, Vietnam has barely any of the usual ground-based infrastructure required to move things about, like trucks or trains.

Instead, Vietnam has barges, hundreds of them. From the Mekong Delta they carry coffee, flowers, coconuts, durians, dragon fruits, cashews, rice, pork, and many other agricultural products. In the hands of the country's remarkably capable street food vendors, home cooks, and chefs, this natural bounty has made the country an internationally famous destination for foodies.

From Vietnam's factories, barges carry things like Hyundai cars, Nike and Adidas sneakers, clothing bearing the labels of Patagonia, the North Face, Gap, Uniqlo, and H&M, as well as the majority of Samsung smartphones sold worldwide, which—surprise!—aren't made in either Samsung's home country of Korea or in what we usually think of as the electronics-manufacturing capital of the world, the Pearl River Delta area of China. Google also shifted manufacture of its phones to Vietnam; Apple suppliers have shifted production of AirPods to the country, and they may eventually make the iPhone there, too.

Many of those barges converge southeast of Saigon at the Port of Cai Mep, on the dredged ship channel in the Thi Vai River. Each barge carries between ten and eighty-two shipping containers, almost all of them forty feet long. In Cai Mep, containers are hoisted by quayside cranes, transferred to a container yard, then moved to the far more massive vessels that will carry those goods to the rest of the world.

Vietnam's major industrial areas are northeast of Saigon, in and around an area called Bien Hoa. A handful of these factories manufacture a great many of a particular item of which you probably own a dozen or two: the USB charger.

Maybe you've never wondered what's inside one of these, or why it doesn't resemble a conventional plug, like the one at the end of any other appliance. Were you to crack one open, you might be surprised to find that in its densely packed innards are the guts of what looks like a tiny computer—capacitors, resistors, diodes, all soldered onto green PCBs (printed circuit boards) crisscrossed with electrical interconnects. And on one side of one of those tiny boards, you would find an integrated circuit—that is, a microchip—no bigger than a lower-case *o*.

The miniature, multilayered tangle of electronics inside a USB charger is necessary to deliver to our gadgets a smooth, predictable stream of five volts of direct current. The complexity inside one of these humble "wall warts" can transform the wide variety of household currents to be found across the globe, which vary by frequency, voltage, and consistency, into exactly what our device needs in order to charge quickly and safely, no matter where we live or travel.

A country with the capacity to manufacture the constituent parts of a USB charger and then assemble them into a finished product is already well on its way to manufacturing much more complex electronics. In other eras, the bellwether of a country's advanced manufacturing capabilities was its ability to make mechanical looms or interchangeable metal parts or transistor radios. Today, it's microchips and the things they control, however simple they may be.

This utterly mundane, deceptively simple object is the one whose path we'll trace from the factory to your front door. Odds are that at some point in your life you've ordered one of these, or a gadget that came with one, and received it the next day. But its journey was in fact at least 14,000 miles and two months long.

In one of hundreds of factories in Vietnam owned by contract manufacturers like Hon Hai Precision Industry Co., better known in the United States as Foxconn, manufacturer of the iPhone, USB chargers pour off the ends of assembly lines. Then they're paired with other gadgets, or individually packaged, and combined in bigger, corrugated cardboard boxes. At the factory loading dock, these boxes are loaded into the outermost shell of this Russian doll of packaging, the one object most responsible for the state of today's world: the forty-foot shipping container.

The modern shipping container, the same kind you've no doubt seen riding piggyback on a purpose-built trailer towed by a semitruck as it rolls down any highway in America, can hold more than 59,000 pounds of cargo. Add in the weight of the steel box itself, another 8,000 pounds, and its total mass is equal to that of some World War II–era Sherman tanks. If it ends up on the bottom of a stack of other shipping containers, it can support 211 tons of additional weight, the equivalent of twenty-seven fully grown male African elephants. On a ship, eight fully loaded containers can be stacked on top of it and even in heavy seas it won't buckle. In the biggest ships, twelve or more of its brethren can be stacked on top of it, as long as some are empty.

The corrugated walls of a shipping container are only 0.075 inches thick. That's less than the thickness of three credit cards. If you walk on top of a shipping container, its flimsy roof will flex under your weight. As with the modern automobile, its tubular steel frame is what makes it strong; its walls are there just to keep things in and the weather out.

A typical shipping container is 40 feet long, 8 feet wide, and 8.5 feet tall. Its interior volume is nearly 2,400 cubic feet. It's basically a 400-square-foot studio apartment stretched out so as to fit on a ship, truck, or railcar.

Lashed to other shipping containers with locks at each of its eight corners, and further buttressed with crisscrossing steel rods against the vertical and lateral acceleration inevitable on an oceangoing vessel, it stays firmly anchored to the ship it's on in even the worst storms. It's watertight. Its 1 ⅛-inch-thick marine plywood floor is infused with chemicals like chromium and arsenic to keep pests away, and its painted walls are similarly infused with toxins. (This is one reason shipping container architecture, far from being some kind of

Earth-friendly "recycling," is in many cases an absurd vanity: stripping away all that nasty stuff takes more time, money, and energy than building a comparable structure anew.)

Shipping containers are what are known as "intermodal" transports, which means they can be carried by two or more means of transport. Intermodality and containerization are quite old. It's a bit obscure at what point the plain old box became an intermodal "container," but as early as the late eighteenth century, wooden boxes of coal were being transported by barge in England, then transferred to horse-drawn wagons.

The shipping container holding our box of boxes, containing our small army of USB chargers, rides on a chassis pulled by a semitruck. This first short leg of its journey is called drayage, and it's one of the most obscure areas of trucking, despite its importance. The hours and wages of drayage drivers can fluctuate wildly and are at the whims of ports, shippers, and the shifting winds of global macroeconomics.

From a factory in Bien Hoa, our shipping container travels by truck to one of two inland freshwater ports, either Binh Duong or Dong Nai. These ports are large for their kind but small compared with the sort that accommodate the giant container ships that cross oceans. Dong Nai alone handled 673,000 twenty-foot-container-equivalent units in 2018, enough to fill one giant oceangoing 10,000-unit container ship every five and a half days.

At Dong Nai, a small port crane resembling a long-necked dinosaur expands the spreader at the end of its steel cable, grabs our container, lifts it from the truck, and transfers it onto a barge.

Most of the shipping containers flowing out of ports in Vietnam hold goods for export. In their teeming masses, they contain the enormous efflux of finished goods from the country, which has become, like so many of its neighbors, one of the world's preeminent workshops. What America's Rust Belt once was, that swath of productive enterprise stretching from Baltimore to the Great Lakes, places like Vietnam have become—only for the whole world.

Seven or so hours later, the barge arrives at the Port of Cai Mep.

Here, a ship-to-shore crane as tall as a thirteen-story building, designed to load and unload massive oceangoing vessels, makes short work of the relatively puny barge. It transfers containers onto the backs of yard trucks, which take each box into the port's container yard.

While many of the world's largest and busiest ports use a great deal of automation, Cai Mep functions in the same way that most ports have for decades. Human-driven cranes, trucks, and—inside the port proper—yet more cranes are used to consume, stack, and sort containers.

This sequence of events, starting with the ingestion of a shipping container into a port from a barge, truck, or train, happens more than 400 million times a year worldwide. That's more than a million times a day, or about 13 times a second. A train long enough to carry a day's worth of the world's shipping containers, stacked two high on each railcar, would stretch more than halfway around Earth's equator.

Nearly 800 million units of container shipping were moved through the world's ports in 2019, according to the World Bank. Confusingly, the unit by which the entire industry measures containerized shipping is the twenty-foot-container equivalent, so every forty-foot container, which is the vast majority of them, counts twice.

The container yards of modern ports are like giant stacks of multicolored children's blocks divided into neat piles, each with a purpose. Carried aloft by a rubber-tired gantry crane, or RTG, our container will be stacked on top of and then rest below others and moved multiple times during its brief stay in the port, as software algorithms dictate the most efficient way to order containers so they will be available when their next destination, a big oceangoing vessel, comes to call.

Tall and spindly, RTGs have four legs, each ending in two or four tires apiece, which allow them to move in only two directions—backward and forward. Spanning from one leg of the crane to the other, across their tops, a frame supports the spreader that can move back and forth to grab a container. Sequestered in the legs of the crane are their diesel and electric power supplies. RTGs look like bridges on wheels, strangely insubstantial compared to the shipping containers they lift.

The RTGs at Cai Mep have the proportions of a pizza box laid on its side. Each is only about as deep as a single container is long, but they are more than sixty feet wide and almost as tall. They are big enough to construct long blocks of containers seven across and five high. In combination with yard trucks and a few other bits of miscellaneous kit, this is how containers are moved about on the otherwise featureless, perfectly flat concrete plain of the port.

Containers typically only stay in the yard at Cai Mep for three to five days. Yet at every decent-size port, these yards are so large they seem like small cities, with boulevards and alleyways hemmed in by four-story buildings made of containers.

At Cai Mep, containers of a dozen different shades, each denoting the shipping company to which they belong, are stacked in long blocks.

Some containers are battleship gray and have a seven-point star on a field of blue next to the word MAERSK. These belong to the world's biggest shipping company, which owns APM Terminals, which is a part owner of CMIT, the operator, under a fifty-year lease from the Vietnamese government, of one of the container terminals at the Port of Cai Mep.

Other, dark blue containers, say APL and have a red eagle on them. In a chain of ownership typical of the ever-more-consolidated shipping industry, APL once stood for American President Lines, founded in 1848 as the Pacific Mail Steamship Company. It's now owned by a Singapore-based company that in 2016 became a subsidiary of CMA CGM, a French shipping conglomerate based in Marseille, formed in 1996 when CMA bought CGM, after the latter was privatized.

Maintaining a fleet of container ships is now so expensive, requiring volumes of capital alien to all but the largest and most complicated forms of human enterprise, that in order to survive, ocean-freight companies have steadily died off or consumed one another. It's a long process of auto-cannibalization that has left the world with just eleven major shipping companies. Even these companies are in some way a fiction, because sitting atop them are just three so-called ocean alliances that allow firms to share ships and containers in order to balance out the variable demand for vessels, and battle one another for market share.

The biggest of those alliances, known as 2M, is the partnership of Maersk and the Swiss-Italian conglomerate MSC. Together, these two giants control just over a quarter of the world's shipping volume. Just behind this alliance is the combination of CMA, CGM, COSCO, OOCL, and Evergreen, which accounts for another quarter of the world's shipping container volume.

To understand the scale of global container shipping—we're not even counting the giant ships that transport oil, iron, grain, and the like—envision the total volume of the more than 5.6 million shipping containers that the world's two biggest shipping alliances can collectively transport.

If you're having trouble, picture every block of Manhattan from 14th Street north to 34th, where the Empire State Building stands now, and from 1st Avenue on the East Side, just a block away from the East River, all the way to 11th Avenue on the West Side, where Hudson Yards begins and the entrance to the Lincoln Tunnel gapes. Now picture every one of those blocks filled with Empire State Building–size towers of shipping containers—205 in all. This steel monolith would contain more than half a billion square feet of floor space, enough to give a million New Yorkers their own studio apartments.

Now recall that we are only counting containers representing the capacity of half of the container ships in the world. To represent the capacity of all the world's 5,000 or so container ships, our collection of Empire State Buildings made of shipping containers would have to stretch more than two miles, from 14th Street to 54th Street—Union Square to the Museum of Modern Art. It would take nearly three hours to circle this mass of shipping container skyscrapers on foot.

And we're not done yet. A careful reader will notice that we are describing how many containers fit on the world's fleet of ships purpose-built for moving them. The total number of shipping containers in use in the world is greater than 17 million, most of which are of the forty-foot variety. This number includes the huge number of containers on shore, most of which are waiting in port or being transported to and fro on trucks and trains. Converted into buildings, this container fleet represents 2,000 additional Empire State Buildings' worth of containers. Our hypothetical city of containerized skyscrapers would now stretch an additional 200 blocks, to 254th Street.

In 2018, the world's container ships moved the equivalent of 152 million twenty-foot containers, according to the United Nations Review of Maritime Transport, which means about four and a half moves for every one of those containers.

Four and a half times a year, a volume of shipping containers equivalent to a grid of 1,200-foot-tall skyscrapers 10 blocks wide and 240 blocks long crosses the world's oceans.

The container terminals at Cai Mep handle but a small fraction of that global volume, but a huge quantity for a country the size of Vietnam—3.7 million units in 2019, or the equivalent of 1.85 million forty-foot shipping containers. In 2015, Cai Mep was operating at about a third of its capacity; by 2019, it was at capacity; and by the end of 2020, with many parts of Asia having contained the coronavirus and exports to the rest of the world booming, it was beyond its rated capacity.

A scant five days after our USB charger was boxed up and packed into a shipping container, that container is plucked by an RTG, laid gently onto the bed of a yard truck, and driven the length of a city block to the underside of a ship-to-shore (STS) crane. Then the driver of the STS crane, peering between his knees, through the plexiglass floor of his cabin, lowers the crane's spreader, locks it onto the four corner posts of our container, and in a matter of seconds lifts it ten stories into the air, while simultaneously swinging it over the deck of the world's largest moving object.

Chapter 3

SHIPS AND OTHER CYBORGS

It's often been said that America runs on trucks, and that if you want to give thanks for the food on your table, the fuel in your automobile, the clothes on your back, and the book or device you're holding, you should thank a trucker. This assertion isn't wrong, but it is myopic. Most of the time, before those goods were loaded onto a truck, they were transported by other means.

At some point in their journey, 90 percent of the world's goods travel by ship. And while things like coal, oil, bauxite, cement, and grain—so-called bulk cargo—accounted for approximately 80 percent of the 11 billion metric tons of goods transported by ship in 2019, the remaining 20 percent was carried inside shipping containers, according to the United Nations Conference on Trade and Development.

Those are difficult numbers to envision. But if, for example, some cosmic practical joker replaced the entire human population of China with adult male African elephants, their total mass would be approximately equal to that of all the material transported by ship in a single year.

Since containers are generally used to transport finished goods, if we look at the monetary value of what's transported in shipping containers every year, these figures are reversed. The "mere" 2 billion tons of goods shipped in containers in 2019, which represent the equivalent of 76 million standard forty-foot shipping containers, constitute 70 percent of the value of all goods transported by ship that year. That's about $4 trillion worth of goods transported by shipping container every year.

If you look around the room you're in now, it's almost certain that all or nearly all of what you can see came by ship.

Most every shipping container in which those things were transported is more or less identical. They are pretty much all constructed from the same materials, with precisely the same dimensions, as dictated by a passel of international standards. The complete interchangeability of all of the world's tens of millions of containers—which collectively flow through the world's ports just shy of 400 million times a year—is the primary reason they're so valuable to the global economy.

If the basis of the internet is a packet of data, the shipping container is its equivalent in the physical world, the discrete unit upon which depends nearly all global exchange of manufactured goods.

One week after I toured Cai Mep in Vietnam, the OOCL *Brussels*—a 1,200-foot-long ship capable of carrying 6,600 forty-foot-long shipping containers, built by Samsung Heavy Industries in one of the largest shipyards on Earth, in Gyeongsangnam-do, South Korea—stopped at the port.

A ship like the *Brussels* is in many ways the standard large transoceanic container ship, and exactly the sort to transport the container holding our USB charger.

Among the world's largest vessels when it was christened in 2013, the *Brussels* is now merely above average, with only about half the cargo capacity of the world's largest container ship. That vessel, the Hyundai Merchant Marine *Algeciras*, is a behemoth as long as the Empire State Building is tall, with a deck so big that four professional soccer games could be played on it at once. Ships comparable in size to the *Algeciras* are too big to squeeze through the Panama Canal, and only the largest ports can accommodate them.

For the foreseeable future, ships like the *Brussels*, so-called Neopanamax vessels, will remain the workhorses of international fleets. At 164 feet wide and 1,200 feet long, it's the biggest a thing can be and still squeeze through the locks of the Panama Canal, which were expanded between 2007 and 2016.

On board the *Brussels* on the day it arrived at Cai Mep, ready to take on thousands of containers of electronics, clothing, PPE, toys, tools, and just about everything else we associate with the "Made in China" label, but which increasingly means "Made Near China," was Third Mate Jeff Tsang. After racking up more than seven years of time at sea, broken up by time ashore to attend maritime university, Jeff, as he likes to be called, was on his final tour on a container ship.

Jeff has black hair, a disarmingly boyish face, and nearly half a million subscribers on YouTube, where he posts videos about life at sea. He was supposed to be on board the *Brussels*, which circumnavigates Earth every seventy-seven days, for just six months. Thanks to the global pandemic, he would stay on the ship for nearly a year.

Under normal circumstances, containerized shipping companies exchange 150,000 sailors every month, pulling as many off ships as they put on, for a total movement of 300,000 people. But when Covid-19 struck, the companies invoked the legal principle of force majeure—extraordinary circumstances—to keep officers, captains, and crew on board their ships until ports allowed sailors to disembark again and cross-border travel on international flights could be worked out.

Then, for the most part, the shipping companies failed to work it out. By the close of 2020, 400,000 sailors, or about one in every four men (and a small number of women) then on a container ship, were trapped at sea beyond the end of their contract.

For some, it was a matter of being kept on for a month or two. But for an unknown number, it meant contracts had doubled in length. And for a minority, it meant not being able to leave their ship for well over a year, a stay far beyond the eleven-month maximum allowed by international law.

It was a crisis so profound, so widespread, and so potentially devastating for crews already worked beyond the point of exhaustion that the International Maritime Organization (IMO), in language calculated to alarm, declared that shipping companies' systematic and ongoing failure to repatriate crewmembers and bring on fresh sailors bordered on "forced labor."

Translating that from the careful, consensus euphemisms adopted by members of the United Nations, what the IMO was almost but not quite saying was that one in four of the world's sailors on container ships was now caught in a modern-day form of slavery.

To be fair, the eleven shipping giants that, after decades of consolidation and bankruptcies, now control nearly all global oceangoing shipping, were at least trying to fix the situation, says Christiaan De Beukelaer, a researcher at the University of Melbourne who studies the plight of sailors. Regulations that precluded sailors from leaving their ships, established early in the pandemic by many countries, weren't, after all, the fault of the companies whose ships were calling on the world's ports.

But in late 2020 and early 2021, as it became clear that crew changes were possible, if expensive, the shipping giants' failure to take responsibility for their own sailors began to look more like cost cutting than unforeseeable circumstances and a lack of action by governments, says De Beukelaer.

"I would encourage each and every one of you to think of how you would feel if you had to work every day for twelve hours, with no weekends, without seeing your loved ones, and trapped at sea," said Captain Hedi Marzougui, a Tunisian-born U.S. citizen, at a United Nations event.

"Not knowing when or if we would be returning home put severe mental strain on my crew and myself," he continued. "We felt like second-class citizens with no input or control over our lives."

At the same event, IMO secretary general Kitack Lim said that on more than 60,000 cargo ships, which throughout the pandemic continued to deliver essential goods, food, and medicine, "overly fatigued and mentally exhausted seafarers are being asked to continue to operate ships . . . [and] safety is hanging in the balance, just as seafarers' lives are being made impossible."

The 1.5 million sailors who are on the world's container ship fleet at any given time had, it seemed, been declared essential workers. So essential that many would not be allowed to stop working at all. Given that the global exchange of goods would grind to a halt without the efforts of workers who constitute just 0.02 percent of the world's population, it was perhaps the second time in history that so much was owed by so many to so few.

▶ ▶ ▶

Jeff, one of the 400,000 sailors caught up in what came to be known as the "crew change crisis," was, in his own way, one of the lucky ones. By dint of life stage, disposition, and a talent for enduring the hardships of modern seamanship, the near doubling of the span of time he would spend on the *Brussels* didn't really bother him.

"When I signed on, I said I wanted to stay on eleven months, and they thought I was crazy, because the standard contract is only six months," he says. In February, after Covid-19 began its malignant spread in China, he got his wish. Some of the crew on his ship ultimately stayed on fourteen months.

Another way Jeff was lucky was that his employer, Hong Kong–based OOCL, which stands for Orient Overseas Container Line, has some of the most favorable working conditions for crewmembers, or at least officers, of any of the major shipping companies. And yet a third way he was lucky was that his home, Hong Kong, happens to be one of the stops the OOCL *Brussels* makes on its regular two-and-a-half-month-long circumnavigation of Earth. This is rare, especially for crew on container ships, many of whom tend to be from the Philippines or Ukraine. Because Jeff didn't have to cross any international borders after leaving the ship, it was much easier for him to get home when the opportunity arose.

The mental and physical challenges of modern seamanship have more in common with the privations of prison or long-duration spaceflight than anything Herman Melville would have recognized as the life of a sailor. The terror of watching a storm sweep over the horizon and threaten every person on board with destruction has been replaced by boredom, routine, and loneliness, but also the exhaustion that comes from interrupted sleep schedules and the kind of high-stakes tedium that characterizes many twenty-first century professions. One wrong move can be career ending for a modern sailor, leading to millions of dollars in damage or lost revenue for their employer.

Jeff's personality, which comes through in his YouTube videos, might be described as monastic, even ascetic. While long stretches away from home affect his fellow crewmen, as someone without a partner or a child, Jeff hardly ever gets homesick, he says. Nor, as an introvert whose passion is making videos, including beautiful, meditative, multiday time-lapses filmed on the ships on which he works, does the lack of shipboard camaraderie bother him much.

Most sailors on modern container ships keep to themselves, retreating to their cabins to sleep, watch videos, or play video games when they're not on duty, says Jeff. Recreation facilities on even the largest and most modern ships are usually limited to a small gym and common area, which probably includes a disused couch or two and a karaoke machine. While shared meals still take place in a galley, alcohol has been banned on all large ships for decades.

It's been a week since our USB charger was loaded onto a truck at the factory, transferred onto a barge, and taken downriver to Cai Mep, where it was sorted into a stack of containers in the port's vast yard. Tomorrow is the day it will be loaded onto Jeff's ship and will leave the country. The day after, on January 30, the World Health Organization will declare a global health emergency.

On this morning, Jeff awakened, as he always does, at around 7:30. He slept soundly, since the *Brussels* has an unusual amenity: the tower that includes the ship's accommodations and bridge is separated from the ship's engines. Typically, sailors must learn to ignore the shuddering of those engines in order to get any rest at all.

By 7:45, Jeff is on the bridge. At 8:00, his watch begins. As one of three officers on the ship, he is on watch for eight hours a day, broken into two four-hour sessions, eight hours apart. This is so an officer is on the bridge and alert at all times. Their job is to navigate, watch for hazards, and monitor the ship itself.

Jeff never eats breakfast. He drinks his sole cup of black coffee at the start of his watch, never before, so as to make the most efficient use of his time.

The bridge of the *Brussels* sits near the front of the vessel, about a third of the way from the bow to its stern. Even though the bridge is more than a hundred feet in the air, the main thing Jeff can see from it is the tops of shipping containers, row after row of them, stacked up to seven high on deck, and with many more belowdecks. These multi-colored rectangles obscure the bow of the ship, but beyond them, he can see the sea, the horizon, and the gigantic dome of sky familiar to those brave, foolish, or desperate enough to make their living as mariners.

The *Brussels* feels more like a building than a ship, says Jeff. Other than calls to port, almost every day on board is the same. Even the parts that at first seem spectacular, like watching the clouds reflect the sun's amber light as the day breaks, quickly become routine. "After a while, you get used to sunsets and sunrise," he says. "When you see it every day for ten months, it's pretty mundane."

When the sea is clear of other vessels, even on radar, which it is for the majority of the distance the ship travels around the world, across the trackless expanse of Earth's two mightiest oceans, Jeff does something unusual.

While many officers on watch stand at the windows of the bridge, staring into the distance, letting their minds go blank, Jeff paces. Back and forth across the hundred-foot-wide expanse of the bridge he walks, his Apple Watch counting his steps. He gets in six miles on every shift.

On the bridge, there is typically only one other person, the helmsman. While Jeff's morning watch is always the same—8:00 a.m. to noon—

the helmsmen rotate shifts, so all of them eventually get to witness his unusual ritual.

While he paces, Jeff uses a speaker or earbuds to listen to audiobooks on self-improvement. The last one he finished was *How to Win Friends and Influence People*, which he enjoyed, even though he found it a bit simplistic. At noon, his watch complete, he hands control of the vessel to the officer with the noon to 4:00 p.m. watch, eats lunch, and takes a one-and-a-half- to two-hour nap. Like everyone else on board, his sleep is fragmented throughout the unceasing twenty-four-hour cycle of the ship, and frequently interrupted.

Early the next morning, on the day the *Brussels* puts in to Cai Mep, and hours before his morning watch would normally begin, Jeff rouses himself to help with the docking process. As third mate, he is liable to have one of two jobs in addition to keeping watch, either looking after the safety of the ship and its crew or managing its cargo. The latter, which is mostly office work, is Jeff's job on this voyage. A week ago, he filled out the ship's "port papers," the Word documents and Excel files that declare who is on board, the ship's gross weight tonnage—its mass, including cargo—and its "flag state," which is the country whose laws the ship has agreed to obey and to which it pays taxes and fees. Since ship owners and operators may flag to any country they choose, a system accurately described as "flags of convenience," carriers favor Panama, which is to oceangoing vessels what the state of Delaware is to corporations in the United States—a low-tax, low-fee, lax regulator chosen by default.

Other than the overtime that occurs when his ship is entering or leaving port, or while it's at the quayside, Jeff maintains his usual watch, from 8:00 a.m. to noon and again from 8:00 p.m. to midnight. Much of what a sailor used to do when a ship is in port is taken care of by various administrative workers on shore, known as the ship's agents. They handle all matters of what cargo is to be put on and taken off, where, as well as myriad other mundane tasks, like procuring the ship's stores of food, fuel, and other supplies.

One of the most important tasks ships' officers used to be responsible for, creating a plan for the cargo to be stowed so that its weight is balanced across the length and breadth of the ship, is now taken care of by software.

Modern cargo ships are so large, and the forces experienced by the containers on board therefore potentially so extreme, that for reasons of both safety and efficiency, some carriers now use sophisticated software to run physical simulations of the behavior of shipping containers on board every vessel at every stop. This software even takes into consideration the weather a ship is predicted to encounter along each leg of its journey. This is important because, as a ship pitches and rolls in a storm, even a very large one that barely budges in twenty feet of swell, its movement subjects the apartment block–size stacks of containers on its deck to g-forces that effectively double the mass of those steel boxes. (This pitching can also do a great deal of damage to the contents of those boxes if they're not packed properly, in a way that fills the entire container.)

The goal of the cargo plan and the software that creates it is to get as much cargo loaded onto a ship as quickly as possible, with a minimum of shifting—that is, moving containers from one place to another rather than onto or off the ship. In the old days of break-bulk shipping, before containers, when pretty much anything might be stowed in the hold of a ship, vessels would spend weeks in port. The many unionized longshoremen who loaded and unloaded ships constituted a powerful political and economic force, especially in the United States, where in 1934 they went on strike for eighty-three days. Labor unrest aside, the inherent unpredictability of the time it might take a ship to be loaded or unloaded meant that goods shipped overseas had to be ordered many months in advance. The speed and predictability of containerized shipping, which started slowly in 1956 and by 1980 represented 90 percent of the transport of finished goods, is largely what made globalization possible.

Even though the OOCL *Brussels*, quayside at Cai Mep, is disgorging and taking on tens of thousands of tons of cargo, spread across thousands of shipping containers, it may stay in port as little as twenty-four hours. This is often not enough time for its crew to disembark. And at many ports, even those who are free to do so may opt to go

ARRIVING TODAY ◀ 34 ▶

to a shop and comfort station in the port itself, where prices for food and other items may be much higher than what they would pay in a nearby city. In an age in which more and more shipping goes through megaports that require vast swaths of land, a trip to a nearby city is often an expensive and long cab ride away.

Less than two days after the ship put in at Cai Mep, Jeff is working overtime again. It's afternoon, and all the containers have been lifted onto the ship, one at a time, by massive ship-to-shore cranes. Each container is bound to the one immediately below and above it by four twistlocks, which are small, spring-loaded contraptions, not much bigger than two fists put together, made from heavy, strong steel and capable of keeping a container secured to the deck or another container even in heavy weather.

Jeff's superior, the chief mate, has handed him the cargo plan. The shipping container holding our hypothetical USB charger has come aboard in much the same way as every other container on every other container ship. Despite being as big as a small apartment, if we zoom out to the scale of the whole planet and all its shipping, our container might as well be a single particle in the vast flow of goods across the widest and most open spaces of Earth.

Because modern container ships are so big and carry so much cargo above deck—as much as a third of it on some ships—the containers that are visible on a ship must have additional lashings to keep them secure. These lashings, which are attached by crew after the containers are in place, consist of long steel bars as thick as a forearm, and on a ship ready to sail, they stretch across many boxes, cross-bracing the ends of containers.

Cargo plan in hand, one of Jeff's jobs is checking that all these lashings are present and secure. Each is tightened with huge, purpose-made crowbars with two-tine forks at the end, which fit into the places where one steel bar screws onto the threads of another.

Jeff must also check that every bay and row of containers, all down the ship's 1,200-foot length, exactly matches the cargo plan. Then he has to check that all of the refrigerated containers on the ship, known as reefers, are both plugged in and powered up. Such containers can maintain a range of temperatures, from cool enough to keep bananas green to cold enough to keep meat frozen.

Reefers now represent 7 percent of all the containers shipped, and their ranks are growing at a steady 5 to 7 percent a year. What was once the sole domain of companies like Chiquita, famous for its fleet of white-painted ships that carried only bananas, is now available to a much wider array of shippers.

An hour before the ship is to leave port, the officer who will take the next watch is roused, as is the captain. The engine room is notified that they should prepare the massive two-stroke engine of the ship to deliver full power.

Deep in the bowels of the ship, in a chamber the dimensions of a midsize medieval church, the walls and ceiling covered in a tangle of pipes as big as sewer mains, one of the world's largest reciprocating engines is brought to life. Diesel engines this size are designed for power but not speed. Modern shipping, in its quest to control costs and emissions, is built around the concept of "slow steaming," where large ships travel at speeds significantly lower than their maximum.

By traveling at eighteen knots (or about twenty-one miles an hour), a container ship crossing the Pacific adds a week to its journey but saves 59 percent on fuel costs—and fuel is the single biggest expense of any trip.

First adopted in the late 2000s when fuel prices spiked, slow steaming has since become the norm across the entire industry. In an industry where time at port is minimized, and the entire reason for containerized shipping is to move things more quickly and efficiently, slow steaming is a paradox. From a shipper's perspective, the savings on fuel matter little, since containerized shipping is already one of the least expensive links in the entire supply chain. For example, it costs about two dollars to ship a TV from China to the United States, port to port.

For shipping companies, which are constantly beset by an over-supply of available ships and capacity, fuel savings matter quite a lot, of course. But another part of the explanation for the paradox of why in a world when everything is speeding up, ships have slowed down is a theme that will become apparent in many other links in the supply chain: in the transportation of goods, predictability matters as much, or even more, than speed.

While every company in every link of this chain is of course focused on doing its job as quickly as possible, when there is an unavoidable trade-off between doing things quickly and doing them in a way that broader systems of people, machines, and software can be sure they can anticipate and count on, it's predictability that wins out. When billions of items must travel to the homes of hundreds of millions of consumers, the sheer scale of the enterprises involved is what dictates this tight linkage between predictability and efficiency.

As a result, an engine like the OOCL *Brussels*'s twelve-cylinder behemoth, which is fifty-six feet tall—nearly the height of a six-story building—is designed for efficiency and reliability above all else.

That's not to say it isn't powerful. At a leisurely 77 rpm, it produces 74,000 horsepower. It's connected directly to the ship's propeller, and the entire power train is designed to spend almost its entire working life running at a single speed optimized both for fuel consumption and wear and tear on the engine.

Since the ship will leave on Jeff's watch, an hour before his shift would normally begin, he's called to the bridge. The ship has been idle at port, so the first thing he does is an inspection of every bit of navigational gear on the bridge. The control panels for the ship's instruments and systems are spread out across a span of a few dozen feet, singly and in clusters, below the giant windows that make up the front of the bridge. He does his inspection systematically, moving from left to right.

The process of readying the nerve center of the largest moving human-made object on Earth is remarkably simple: Jeff just switches each instrument on and looks for any alerts. Each system is so complicated that the only way to make it reliable is to build in all of its own diagnostic routines.

Fifteen minutes before the ship is to leave port, the harbor pilot, also known as the marine pilot, comes on board. Wherever a captain takes his ship in the world, once it's within a few miles of a port, it is no longer his own. Unique local conditions and navigational hazards, tight squeezes between the sides and bottoms of ever-larger ships and the channels that must be continuously dredged and expanded to accommodate them, as well as tradition, all dictate that ships be taken into and out of port by seasoned harbor pilots for whom this is their sole job.

Once the *Brussels* is back on the open ocean, the harbor pilot leaves just as he arrived when the ship first approached Cai Mep: on a fast, tugboat-size ship used solely for ferrying pilots. The captain is once again in control of his ship.

For Jeff, who is in some ways just a human fail-safe for automated systems that usually require no intervention, leaving Cai Mep is a change of tempo. It's also the beginning of the most dangerous and hair-raising part of his regular circuit around the world.

"There are no other places in the world that are busier than the South China Sea," says Jeff. Here, the captain and every officer on watch must all be on high alert.

Fishing boats, cargo ships of every size—from rusting old bulk carriers to the largest container ships afloat, "Post-Panamax" vessels carrying nearly twice the cargo on the *Brussels*—and assorted other flotsam litter the ship's radar, the horizon, and all the sea between the sky and the bow of the ship.

The navigational technology on board a modern container ship includes two kinds of GPS, one to access the U.S. network and the other the Russian; the Automatic Identification System (AIS), on which ships constantly broadcast their identity, heading, and speed; the Electronic Chart Display and Information System (ECDIS); and a succession of backup navigational aids from every era of sailing between the twentieth century (the electronic gyrocompass) and second-century Rome. Despite all that gear, sailors in highly trafficked waters must rely on their own eyes as a first line of reconnaissance.

Partly this is because many ships below a certain size do not have things like AIS to make plain where they are and where they'll be in the near future. This means that, aside from showing up on radar, these ships are invisible to the electronic systems designed to connect ships to one another and to shore.

Hence, on watch in the South China Sea, Jeff is constantly scanning the horizon with an old-fashioned pair of binoculars. As soon as he spots a ship—be it an unregistered derelict fishing vessel with unlicensed crew who are likely to be refugees conscripted into forced labor by unscrupulous employment agents, or a barge, or a pleasure craft—Jeff switches his attention to the radar, identifies which blip on the screen corresponds to the ship he's spotted, and lets the onboard computer do the work of calculating the ship's CPA, or closest point of approach, and the time it will occur.

Like an astronomer examining the trajectory of an asteroid to determine whether it will strike Earth, this procedure allows Jeff to determine whether there is any danger of colliding with another ship. Oftentimes, it will appear that a ship is on course to ram the huge target of his 1,200-foot-long vessel, but the computer and radar allow him to assess that it will pass just ahead or astern of his ship.

The challenge in relying on these automated systems is that, as in traffic on a busy road, every vessel is reacting to those in its path while also trying to save precious fuel and time by minimizing deviations from its course. And unlike a busy intersection, no one can make eye contact, wave the other person on, or flash their headlights. Communication through medium and long-range VHF and UHF radio is possible, but not always.

The main reason vessels large and small aren't constantly crashing into one another in the world's most crowded seas—which include this one, the approaches to the Panama and Suez Canals, the Strait of Gibraltar, and the English Channel—is a globally recognized system of rules of the road known as the International Regulations for Preventing Collisions at Sea, or COLREGs.

COLREGs include some common-sense rules—vessels should change speed and heading as infrequently as possible and to a degree that makes their change obvious to other ships, so they can react accordingly—and a few that are necessarily arbitrary. For example, when the paths of two ships cross, the one on the port (left) side must always yield.

In rare circumstances, no matter how many regulations have been established and how clearly a ship's captain and officers perceive the situation, crews have to improvise.

In the early months of 2020, mariners noticed something peculiar in the South China Sea. Many nights, on the approach to the Chinese mainland, an oddly straight line of blips stretched from one edge of ships' radar screens to the other, bisecting the sea like a line of tracer bullets fired into a night sky.

The first time they witnessed it, Jeff, who was on watch, and the captain, who was on the bridge because this was such an unusual situation, weren't worried. Usually, when a large ship approaches a cluster of smaller ones, the smaller ones sort themselves out, since, after all, they're much more maneuverable, Jeff later recalled in a video he made about that night.

"As we get closer to the group, we noticed there was something strange. The silhouettes were too big to be fishing boats. And that's when we realized they weren't fifty-meter fishing boats at all. They were hundred-meter barges. All of them. More than two hundred of them in a line, and we can't see the end of the line."

The *Brussels*'s electronic ship tracking systems were overloaded by the density of vessels on radar. Through binoculars, these barges— massive, oddly fast, with more than enough momentum to do horrendous damage to a container ship like the *Brussels*—seemed endless.

Worse yet, the barges were traveling so close together that there were no gaps for the *Brussels*, or any other large ship, to pass through. Worse, says Jeff, mariners know that Chinese ships often don't bother following international collision regulations.

Looking out at what was effectively an impenetrable wall sprung up in the ocean, the captain said he'd never seen anything like it in all his forty years of sailing.

The crew of the *Brussels* couldn't just stop dead in the water and wait for the barges to pass—at their speed of twelve knots, it could mean waiting for hours. Ships that are late to their scheduled arrivals in ports can create a cascading sequence of expensive logistical nightmares and millions in losses for the companies that own or lease them, including fees for late delivery and pickup from shippers. Because the short-hop journeys of large ships are so tightly scheduled, this can also mean delays and more losses at every port following the one they are late to.

Eventually, the captain decided to gamble. Aiming for one of the too-small gaps between a couple of the barges, he lowered the ship's speed to a minimum and committed to his course.

"The worst that we feared happened," Jeff continued. One barge shot ahead of the *Brussels* so close to its bow that they couldn't see it over the stacks of containers that create a blind spot there. "Our palms were sweaty, knees weak, arms heavy. We could only pray," he said. "Right at that moment, the bridge was dead silent."

The crew waited for the crunch of tens of thousands of tons of steel on steel, the crack of welded seams bursting, the shudder of a mighty vessel catching an obstacle big enough to put it in dry dock, dragging it through the water, both ships helpless, both captains finished.

Then they saw the bow lights of the barge ahead peek out on the starboard side of their ship. The barge had missed by just 170 feet—one-seventh the length of the *Brussels*, and so close that it rocked in the bow wave of the big ship as it passed by.

Astern, things had been even closer—recklessly, the barge following had crossed just a hundred feet behind the container ship.

Hours later, still in darkness, they cruised into the port of Hong Kong by night. A sleek, modern jewel of skyscrapers and city lights emerging from a verdant riot of trees, Hong Kong is, according to Jeff, one of the most beautiful ports on Earth.

▶ ▶ ▶

After stopping to unload and take on containers at three more ports in China—Yantian, Xiamen, and finally Shanghai, the *Brussels* begins its twenty-day journey across the Pacific.

At one point on this journey, its bulk is perched 1.2 miles higher above the surface of Earth than the peak of Mt. Everest towers above sea level. But to those of us who think infrequently of the oceans, this demonstration of Archimedes' principle, this leap across the deepest known chasm in the solar system, is but a ship afloat on the sea, and not, as we might otherwise recognize it, a feat equivalent to flying 700 Boeing 747s' worth of cargo over the Himalayas.

Crossing the world's largest ocean after navigating the hazards of the South China Sea is a study in contrasts. Here, the sea is so endless, so thinly populated—and by only the largest vessels—that an officer on watch can go days without seeing another ship on radar, or weeks without seeing one with his own eyes.

Here, on the swells of the Pacific, the way in which modern container ships have much of the intelligence required to sail them built right into the ship itself becomes apparent.

Take navigation. On ships of old, sailors steered by the stars and, if they were lucky, the astrolabe and superaccurate marine clocks. Holding a ship on a steady heading was an actual job; the helmsman had to gently turn the big wheel connected to a ship's keel while watching his heading on a compass, to account for small deviations in the ship's path.

Nowadays, ships still have steering wheels, but they look like toys. They're plastic, and smaller than the steering wheel on an automobile. They're also rarely used, unless a ship is coming into port and tight maneuvers are required.

No, modern container ships, each the size of a skyscraper toppled onto its side, pushed through the sea by propellers nearly three stories tall, directed by a rudder of the same dimensions, are steered by a dial about the size of the volume knob on a stereo amplifier.

Spin this little knob, and an LED readout shows the heading you've just dialed in. If you want to change course, you spin it a little more. This is the ship's autopilot.

In an age in which fully self-driving vehicles still have yet to arrive, it may surprise you to learn that the first autopilot systems on ships date back to the age of the *Titanic*. That system, known to sailors as the "iron mike," harnessed a gyrocompass—a device still in use today that uses the interaction of a spinning gyro with Earth's gravity to always point to true north—to a straightforward electromechanical device that would apply small corrections to a ship's rudder in order to keep the vessel on a programmed heading.

What's remarkable is how little, in basic principle of operation, the modern ship's autopilot has changed.

Many other features of the *Brussels* and its systems make its operation much simpler, and its maintenance much less taxing, than in times past. The ship's sheer bulk makes it proof against rough seas. And even if it should encounter them, its hull is sealed in a way that allows it to tolerate even sixty-five-foot-high waves and a 40-degree roll. The ship's engine is designed not just for efficiency but ease of maintenance—cranes within the engine room allow crew to remove and replace substantial portions of it even while at sea. And while the outside of the ship must have its paint continually chipped away and replaced, the paint below the waterline includes antifouling agents to prevent a buildup of barnacles and other biological matter.

Centuries of experience, craft, and technology accumulated by sailors and naval architects are the reason the total crew of even the largest container ships is today between twenty and thirty people. In a way, the ship is a giant suit worn by its humans. In their fusion of machines and their own minds and bodies, sailors, like most all of us, have become cyborgs.

Even as a vessel on the transpacific route approaches the western coast of the United States, traffic remains thin. On the bridge, for eight hours a day, from 8:00 to noon and 8:00 to midnight, Jeff paces, getting in his six miles per watch, listening to his self-help books.

He has learned that out on the ocean, with no access to the internet during working hours, he has ample time to think. He ponders how to make the best use of his time, how to apply the lessons he learns from his books, and how to cultivate the self-discipline he prizes.

On a ship like the *Brussels*, in a container indistinguishable from the millions of other steel boxes just like it, on a pallet, inside a cardboard box, individually wrapped, is the USB charger that, despite traveling more miles in the past thirty days from Vietnam to the territorial waters of the United States than it will travel in the remainder of its trip, has only just begun its journey to our front door.

Chapter 4

COMING TO AMERICA

←————————————————————————————————————

By Saturday, February 29, or thereabouts, the container ship carrying our USB charger is sidling up to the dock at the Port of Los Angeles. On the same day, the United States will report what is believed at the time to be the first death from coronavirus on American soil.

My own visit to the Port of Los Angeles begins at 2:00 a.m. on a weekday not long before the arrival of our hypothetical USB charger.

From the west, atop a hill in San Pedro, the port is unmistakable. Above Terminal Island, which makes up the bulk of the port, the light of innumerable sodium-vapor lamps turn the sky a distinctive, sickly shade of orange. Even this early in the morning, from miles away, it's possible to see dozens of cranes, thousands of stacked shipping containers, and a handful of massive ships so still beside their quays they seem more like permanent fixtures than floating visitors.

San Pedro is the suburban Los Angeles neighborhood where many of the port's longshoremen live, a place as steeped in the culture of ships and shipping as the port itself. It's a neighborhood both quaint and quintessentially L.A.—unassuming but world-class taco trucks, transplanted palm trees, and a steady ocean breeze. The houses are small and crowd one another, but the sidewalks are clean and the lawns manicured. People here have good, union jobs.

Even at this remove, there is a faint clanking, a hum like tires on the highway, a sense that in the distance, some gargantuan machine is in motion. Already, overhead gantry cranes are picking up, moving and stacking shipping containers, preparing the yard for the thousands of "moves" of these forty-foot, multiton boxes, both out of the port and onto trucks, and into it from trucks to the yard.

This morning, as on most mornings, a tidal wave of cargo is on its way. It will pour off some of the world's biggest container ships—that is, the world's biggest moving objects.

Everything about the next twenty-four hours is about getting those ships into port safely and then unloading and reloading them as quickly as possible. Accidents can and do happen, gumming up the works, putting trucks and cranes and entire ships out of commission, upsetting shipping schedules plotted months ago, leading to cascading failures in systems that depend on each vessel getting in and out in its allotted time, usually around twenty-four hours. From before dawn one day until the same time the next, containers must be removed and replaced, and possibly resorted, as the weight and nature of cargo are rebalanced across the hold of a ship to ensure the heaviest goods are on the bottom and that their weight is evenly distributed port and starboard, forward and aft.

Everything about the next twenty-four hours, in other words, is about not fucking up.

By 3:00 a.m. I'm standing at the southern tip of a half-mile-long artificial pier, one of the oldest at the port. Here is the historic Municipal Warehouse No. 1, the port's first bonded warehouse, a place where goods can rest and be sorted before customs duties are paid. Built in 1917, it's now empty and disused. Next to it is my destination: the nerve center of the port.

From the mountain lion–infested, smog-entrapping hills of the Angeles National Forest on its northern border, L.A. and its surrounds descend like the wattle on the throat of a turkey. At the nadir of that peculiar geography, dipping into the Pacific Ocean in a way that allows visitors to witness both sunup and sundown over the sparkling waters of the Pacific, stands the Los Angeles Pilot Station. A plain, two-story office building, its most distinctive feature is that it has commanding views of not only the entire port but also all the ocean stretching between the city and the horizon.

Nearly every port in the world has, by law or custom, a public or private service that provides local pilots fit to navigate ships large and small into a port.

What might seem like an outmoded convention—you want to come to our port, you have to pay local pilots to bring your ship into it—is in fact an absolute necessity. Even the ancient Greeks and Romans employed such pilots, men with detailed knowledge of local geography, winds, currents, tides, shoals, shipwrecks, and navigational conventions, to bring their oceangoing vessels into port.

No amount of advanced technology has obviated the need for such pilots. Today's ships are so immense they are built right up to the limit of what ports can accommodate, and sometimes beyond, frequently requiring governments to invest billions in more dredging, higher bridges, and bigger berths than ever. When ships of this scale enter a port, piloting them is a bit like playing the children's game Operation, only the consequences for accidentally touching channels, ships, bridges, or the ocean bottom are catastrophic.

As a result, today's harbor pilots are more skilled, and more necessary, than ever. They are also compensated accordingly: harbor pilots at the Port of Los Angeles, who are municipal employees, make, on average, $434,000 a year. Salaries for harbor pilots elsewhere in the United States are, on average, nearly as much.

Today, I'm shadowing a pair of harbor pilots on a journey both utterly routine and strangely perilous. Despite happening a thousand times a day all across the globe, despite myriad safety precautions, if you're a harbor pilot, doing your job can kill you.

▶ ▶ ▶

Last night, a hundred miles offshore, a floating building hove into view on shore radar. Seconds later, through a VHF-band system known as the Automatic Identification System (AIS), everything salient about the ship—its name, destination, tonnage, speed, and heading—is made known to the on-duty harbormaster at the Marine Exchange of Southern California, the nonprofit body that coordinates all traffic coming into the combined ports of Los Angeles and Long Beach.

Like twin cities, these two ports abut one another, separated in some places by no more than a fence bisecting Terminal Island. They are inextricably linked, both physically and historically, being the products of fierce battles over land and revenue fought in the first decade of the twentieth century. The ports of Los Angeles and Long Beach have distinct traditions, histories, fees, funding, and governing bodies, but geographically, and in many respects economically, they are a single port. Together they are far and away the largest port for containerized freight in the United States, handling nearly 17 million units a year, or the equivalent of 8.5 million forty-foot containers.

The port of L.A./Long Beach is the entry point for half of all the goods shipped to America from Asia. No other port even comes close; in second place is the Port of New York and New Jersey, which handles only a fifth of the volume of goods coming from Asia.

Coordinating traffic in and out of America's busiest container port is no small thing, and it must be accomplished by a single authority, the Marine Exchange. It was once but a shack high on a hill. It's still high on the hills overlooking the port, but it's now, more than anything, an IT hub for all the data that must be fed to the entire port. If the port were an airport, the Marine Exchange would be its control tower.

That offshore behemoth that just popped up on both radar and AIS is the Cosco *Netherlands*. Despite its name, it is a Chinese-built, Chinese-owned ship, flagged to Hong Kong and a part of Cosco Shipping Lines, the parent company of which owns the third largest container shipping fleet in the world, with 507 container ships and an additional 773 ships for bulk cargo. This fleet, collectively the largest in the world, can carry more than 100 million tons of deadweight, which describes the maximum amount of cargo, crew, fuel, freshwater, ballast, and provisions a ship can carry.

A hundred million metric tons of material may sound like a lot, and in some ways it is. In terms of sheer scale, it's on the same order of mass as a small asteroid such as Itokawa, which has a diameter of 1,100 feet but a low density as far as space rocks go, since it's basically a big aggregate of nonmetallic dust particles. But 100 million tons is also only about two days' worth of the total tonnage of goods shipped throughout the United States by all means of transport, or three days' worth of what we ship by truck. In part that's because the United States ships enormous quantities of very heavy things, like grain, coal, oil and its refined products, etc., whereas much of what Cosco ships are carrying are finished goods of relatively high value. A ship full of containers destined for America's retail outlets and a ship full of crude oil are comparable only from the point of view of a naval architect.

The *Netherlands* may be carrying at one time only a tiny fraction of Cosco's annual tonnage, but it is a huge amount of material, on a human scale.

At 1,200 feet long, 170 feet wide, and 157,000 deadweight tons, this ship is one of the largest that call at the Port of Los Angeles, says Craig Flinn, chief of the port's harbor pilots.

Craig is my guide for the morning. He's voluble, affable, six-foot-something, with white hair, a vicelike grip, an encyclopedic knowledge of ships and their navigation, and a Five Boroughs brogue. The first thing he does is offer me coffee. He is also drinking coffee. Everyone I meet in this place at this hour is drinking coffee—it's 4:00 a.m.

Craig is, by all appearances, just happy to be here. He is a man of the sea who has never left it and now has the most desirable job in all the merchant marine. He gets to pilot ships, but he also gets to go home when he's done. For him, there is no shipboard life, no calls at foreign ports, no lonely nights at sea.

Craig and I meet inside the harbor pilots' control room. There are two wide desks facing the bay windows framing a view of the port, which at this hour is still just endless lights against an orange-black sky. For someone sitting at either desk, the view is almost entirely obscured by a handful of large monitors. Most of these displays are, in one way or another, digital facsimiles of the old paper sea charts that were once the sole means of depicting and navigating the port and all its features.

One of the displays looks like Google Maps and is centered on the ocean just south of the port; another is a radar display; yet a third is an inversion of a typical map, the ocean and land an inky black, with ships standing out either as vectors, conveying their heading and speed, or small empty circles, depicting vessels at anchor. Sitting at one of these desks is Beth Adamik, the dispatcher on duty. It's her job to understand and monitor, using her own systems as well as data passed on by the Marine Exchange, where each ship is and when it's scheduled to arrive. When ships get within a few miles of the port, Beth hustles pilots out the door.

Twenty-four hours ago, an agent for the *Netherlands* called the harbor pilots' office to order a pilot. At roughly the same time, the agent also called the private tugboat firms whose nimble, powerful vessels are essential to maneuvering a Goliath like the *Netherlands* up a narrow, man-made channel and then parallel parking it at a berth where clearance between itself and the ships off its bow and stern may be as little as a tenth of the length of the ship.

At 4:30 a.m., the radio at the control station beeps. Beth responds, "Cosco *Netherlands*, this is Los Angeles pilots, good morning."

Little else is said or needs to be. Thanks to the ship's automatic navigation systems, its pilot and officers know precisely where it is on Earth's surface, even in conditions of zero visibility, such as darkness, heavy fog, or driving rain. And thanks to the ship's constant broadcast of its location and heading, through its AIS system, the Marine Exchange and the harbor pilots know precisely where it is and what it's doing. This information is verified by radar.

Navigational and positional redundancy is everywhere in these systems, with multiple pairs of eyes monitoring them all. It's a wonder ships coming into port ever collide anymore, but they do, almost exclusively on account of bad judgment rather than bad information.

Captain John Betz enters the control room. While Craig is my guide, John is the person who will actually be piloting the *Netherlands* past the many potentially journey-ending hazards in its path. John will take it from miles out at sea to within mere inches of where it needs to be alongside the pier, like an elephant finding its precise mark on a stage barely big enough to accommodate it.

John is tall and white-haired like Craig, but lankier in a way that telegraphs a different sort of athleticism.

▶ ▶ ▶

Thirty minutes later, we're headed to the pilot boat, not a hundred yards from the pilot station. It looks like a police cruiser or a Navy PT boat, all gunmetal gray and stainless steel on top. It's too dark for me to see the crimson color of her hull, intended to make her look distinct from all other ships in the harbor. She's only about fifty-five feet long, but powerful, a workaday speedboat able to maneuver alongside vessels big enough to make waves of their own. Silent behind the wheel are a pair of boat handlers: pilots who convey pilots, and safeguard their lives.

I have to climb down a short ladder to get into the boat, the cold steel of its polished rungs slick with condensation. Instinctively, I grip harder, realizing that it would be easy enough, were I distracted or trying to juggle my voice recorder, to fall flat onto the unforgiving steel deck beneath me. It's a harbinger of what's to come for the pilot we're conveying.

As we pull away from the dock, John and Craig explain what makes the geography of this port so distinct. To port—that is, to our left, or east, as we're heading south, is Terminal Island. It's mostly landfill, built up at the beginning of the twentieth century, and shared with the Port of Long Beach. Underneath us is the dredged channel that separates the neighborhood of San Pedro from Terminal Island. Before it was America's largest port, Terminal Island was mudflats made up of millennia of silt dumped in the harbor by the Los Angeles River and its lesser-known companion, the Dominguez Channel.

L.A.'s rivers are one of those things like its subway, the existence of which surprises outsiders and even Angelenos themselves. Yet they are no less important to the city's growth and development than the Thames is to the city of London.

As we pick up speed, skimming across surprisingly calm water, I'm reminded of the opening lines from *Heart of Darkness*: "The old river in its broad reach rested unruffled at the decline of day, after ages of good service done to the race that peopled its banks, spread out in the tranquil dignity of a waterway leading to the uttermost ends of the earth."

We're in the half-sheltered cabin of the boat, but the wind is picking up, and soon we are climbing ocean swells. Here is a thing few ever experience: the full force of the open ocean in a small boat. Craig tells me this is a relatively calm day, and I can't imagine how anyone maintains their footing on bad ones, cutting across waves taller than a person, at twenty-two knots, rigid hull slamming into one crest after another with the force of a steam hammer.

Turning around, we can see the port clearly now for the first time, all of it, broad and flat against the still-dark sky, its natural advantages compared with other West Coast ports now obvious.

The port in Portland, Oregon, challenges ships with a long river leading to its berths. Oakland is surrounded by mountains that make getting containers in and out by rail difficult. But L.A. is flat all the way down to the port, with a good harbor carved, over millennia, by the steady outflow of the river itself.

San Francisco's shipping lanes "shoal up," which means silt on the ocean floor doesn't stay where you put it, requiring frequent dredging and giving rise to the environmental nightmare of where to put all that dredged material. Southern California being a desert, its rivers barely flowing at all, L.A.'s port doesn't have that problem.

"We like to take credit for our heightened efficiency, but the reality is we could just be enjoying great geography," Craig says, laughing.

Fifteen minutes later, we're coming up on a cluster of lights just above the horizon. The *Netherlands* doesn't so much approach as loom. The closer we get the more it towers, until we are near enough that seeing the top of its bridge requires us to crane our necks. Suddenly, a searchlight on our boat pops on, bathing the *Netherlands* in a brilliant white. The name of the ship, painted on its side, is two stories tall.

"If you step outside, you can get a good feel for the scale," says Craig. We exit the shelter of the boat's cabin and are confronted with a sheer cliff of steel, its royal blue paint so uniform it looks like enamel. Below it, water flows between our two vessels like an angry river at flood stage, the chop and churn of the ocean now loud enough to drown out the drone of our engine. Ocean spray flies in our faces; the overpowering smell of the sea fills our nostrils. In the face of this gale, everyone has to shout to be heard.

"I mean, we're looking at a building here," says Craig, and as I look up past the top of the hull, to eight-foot-tall shipping containers stacked seven high, I have the impression of being on the sidewalk in midtown Manhattan, at the bottom of the canyon formed by so many offices and apartments.

We circle the ship, coming around her stern so as not to be plowed under by her bow as she cruises at about eight knots. We see her massive rudder, big as a castle gate, then cut across her wake and come up the lee side of the ship. Vessels this big are like islands, and they can shield smaller ones, like ours, from wind and waves, creating a small, temporary harbor in their hydrodynamic shadow.

Halfway up this side of the ship, between her midline and her bow, a small square of light appears. As we close with the ship, the square gets bigger, and now it's apparent that it's ten feet on a side. Two men in hard hats stand inside this portal, framed by its light, watching us. Seeing humans in the midst of the flank of an otherwise dark ship is jarring. Everything about this scene evokes the only reference point I have for things of this scale and appearance—that moment in movies when a shuttle approaches a space station or interstellar vessel and a long tracking shot takes us across its surface, conveying its awe-inspiring bigness.

Finally, we're alongside, and the most dangerous moment of this entire journey—both ours and that of the little USB charger, somewhere aboard this container ship or one just like it—is nearly here. While the more experienced boat handler stays behind the wheel, John and a body man walk out of the cabin of the pilot boat and up to its bow.

From the portal in the side of the *Netherlands*, a Jacob's ladder dangles. It is literally, still, after all these centuries, just a rope ladder with wooden rungs.

Everything that happens now is about perfect timing. The pilot at the helm of our boat must come alongside at just the right moment. Both our boat and the massive ship John must board are pitching independently of one another, like children on opposite ends of a seesaw. From the point of view of the harbor pilot, this means the ladder he must mount is moving up and down before his eyes as much as ten feet. From the moment he grabs hold of it, it's a carnival ride where the only thing keeping him on is the strength of his grip.

"I've been chased up the ladder by the boat," says Craig. "Right at the time I got on there, all of a sudden the swell came through and up came the boat—so I started scurrying up the ladder like a monkey."

If John steps from pilot boat to ladder at the wrong moment, he could be forced to do what Craig did on that day. If he's not fast enough, the rolling of the larger vessel could dunk him in the water.

"The percentage of survival is minimal if you go in the water in heavy seas, even with a vest," says Craig.

Going in the water exposes the pilot to all manner of hazards: He could be crushed between the ship and the pilot boat. Or the force of the current between our vessels—both are doing about eight knots, which makes them steadier and easier to control—could tear him from the rungs of the ladder. Then the pilot boat might run him over. Or, if his body is sucked toward the ship, he'd go through its spinning thirty-foot-diameter propeller.

Every precaution is taken, and everyone is trained and retrained regularly, but across careers that span decades and many hundreds of boardings, the average harbor pilot has a one in twenty chance of dying on the job.

John and his body man pause at the bow, the deckhand gripping John's life jacket at the back, near the top, where there's a handle attached for just this purpose. Both watch the buck and heave of the larger vessel, visually marking the lowest rung of the pilot ladder as it drops into the violently frothing ocean. John counts up five rungs from the ladder's lowest ebb—marking his target. At precisely the right moment, when that rung is shooting up toward him, the pilot boat just a foot from the ship, John takes one long, graceful step, plants a foot on the ladder, grabs another rung with both hands, and hangs on for dear life.

Craig has done this hundreds of times. One day his teenage son, with whom he surfs, asked him why the side of his surfboard had a dent in it. Until that moment, Craig had never realized how hard he grips his board when he feels a wave coming up, a habit he can't shake.

As soon as John is on the ladder, he's climbing, swiftly but surely. Even now he could still fall off rungs slick with ocean spray. On December 30, 2019, in New York Harbor, Dennis Sherwood, a harbor pilot with thirty-five years of experience, fell from a Jacob's ladder while boarding the Maersk *Kensington*, a container ship much like the one John is boarding now. He fell to the deck of the pilot boat below and died soon after.

Today, John is, as usual, both lucky and good. The entire sequence, from stepping off our pilot boat to pulling himself into the bright portal in the side of the *Netherlands*, takes just a few seconds. Our boat accelerates, veers off, and in the noise and wind and what feels like a cloud of residual adrenaline, the massive vessel disappears back into the darkness.

Chapter 5

PARALLEL PARKING
1,200 FEET OF SHIP

On board the Cosco *Netherlands,* John greets two deckhands, and the three of them take an elevator barely large enough to accommodate them all up to the ship's bridge. Here, a wall of glass looks out on the ocean from a height equivalent to the penthouse of a sixteen-story building. The ship is now three miles out from the entrance to the port.

The first thing a ship's mate hands John is a "pilot card"—a one-page sheet that tells him everything particular about this ship, all of it critical to his ability to guide it successfully. It's like hopping into a rental car, except John has to create a complete mental model of how it's going to handle rather than taking it for a spin.

The card tells him the ship's engine speeds; the horsepower of her bow thrusters, which are used for maneuvering in port; her current draft fore and aft; and her "air draft," which means her height above the water, critical for slipping under bridges and cranes. It also alerts him to the peculiarities of how the ship handles: Does she make wide turns? Is she underpowered for her size? Is she carrying close to her maximum weight, which would mean she'll take longer to stop, or did she drop off so much at her last port of call that she's riding high in the water? Craig Flinn calls the pilot card the "bingo sheet for every detail a pilot needs."

This is all part of the master-pilot exchange, a handover of control of the ship that is both ritualistic (there can be only one commander on the bridge at a time, and his word is law) and practical (handing over a ship also requires the rapid transmission of all pertinent information).

After looking over the pilot card, John goes through a verbal checklist of critical questions: Is everything on the ship working, or are there any deficiencies or issues? Are the anchors manned, clear, and ready to be dropped? Did the captain post lookouts? And on and on.

Then he turns his attention to the ship itself, taking in its course and speed. He's using the view out the windows of the bridge, the ship's instruments, and the navigation gear he's brought on board to understand where the vessel is and where it's headed.

In the days of sail, John might have physically taken the wheel of the ship. But these days, he's not going to touch the ship's controls at all. He will direct every movement of the ship that follows, from the moment he takes command until the moment she is safe at her berth, through verbal commands to the helmsman and officers on the bridge. "Half ahead, steady as she goes," he might say, and the helmsman will respond by setting the ship's engine to transmit half power to the ship's propeller, while making slight adjustments to the angle of the ship's wheel, and therefore the pitch of its rudder, in order to maintain its current course. Over radio, John gives commands to the tugs that will attach to the ship, telling them where to attach with ropes or push on the hull of the ship, how hard, and when to just drift at the end of their tethers, deadweight useful for slowing the ship as it completes turns.

John issues his first command: a slight course correction, which he indicates by saying the compass bearing he wants, and a drop in speed. On the radio, he calls the harbor pilot dispatcher, Beth Adamik, and gets an update on all traffic moving inside the port.

Twenty minutes later, the ship is approaching the entrance to the Port of Los Angeles. It's a thousand-foot opening in the thin, rock breakwater that guards the port from waves. To port, that is, west of the ship, he looks down on the squat Los Angeles Harbor Light, a lighthouse that has alerted mariners to the port's entrance since

1913. While it's no longer required for navigation, it still shines a 217,000-candela beam out to sea, which is only about three or four times brighter than a modern car's headlights.

Like any harbor pilot worth his salt, John doesn't trust the ship's onboard positioning system. Too often, the people who install the antennas that are supposed to let pilots and the rest of the world know a ship's precise location forget or do not care that they may need to be calibrated to reflect that they aren't on the centerline of the ship.

As a result, John has brought his own GPS system and navigational beacons, and he affixes one within the bridge and another outside the bridge, on the railing of one of its open-air "wings." He pulls from his pilot's bag a Portable Pilot Unit, which plugs into the ship's Automated Identification System and begins receiving signals from both the United States' and Russia's global positioning systems. Triangulating from both simultaneously helps him know, on a second-by-second basis, where the ship is at the resolution of a meter or less.

In lieu of a paper chart, John takes out an iPad, which is fed data from both the antennas he just installed and the ship's computer, which passes on data about the vessel's heading, as determined in relation to the ship's gyrocompass. The app he's running shows him a sea chart with an outline of the ship he's on at its center. Ahead of the ship is a translucent gray image of where the ship will be in six minutes, based on its speed, heading, and rate of turn.

Beyond the lighthouse and inside the breakwater, three powerful tugboats are waiting for him. He radios each in turn to tell them where to "make up" on the ship, that is, where to attach themselves with ropes. Two attach to the ship on its port side, since it will have to make a sharp left turn once it arrives in the constricted space of the aptly named "turning basin," deep within the port. One tug attaches to the stern of the ship to drag on it from behind and slow it within the narrow ship channel.

The forces involved in changing the course of a vessel with a total mass close to 200,000 tons are phenomenal. Before the advent of modern towropes, attaching a tug to a ship meant putting such terrific strain on the cables involved that when the ropes or steel cables snapped—and they did, often—they moved with such speed and force that they could kill. Heavy rope or steel cable might not seem like it has much elasticity, but at these scales it's like a rubber band stretched between your fingers, almost to the point of breaking.

US Navy Presents: Synthetic Line Snapback is a training video from the early 1980s that's meant to be serious but, delightfully if you're of a certain age, is in the style of an after-school special. The film warns that when a rope breaks "at 700 feet per second, the parted line snaps back at nearly the speed of a .45-caliber bullet." In slow motion, dummies on the deck of a Navy vessel are obliterated by snapping cables as if by a cannonball. We hear from a seaman who lost both legs in just such an accident, and other subjects describe watching separated ropes kill fellow crewmen.

That this doesn't happen nearly as often these days is a miracle of modern chemistry, and due entirely to the exploitation of forces at the atomic scale. Today's towropes are made of high-modulus polyethylene (HMPE), a plastic made up of extraordinarily long molecules. These molecules bond tightly to one another in such a way that ropes braided from fibers made of HMPE have almost no stretch, even under tremendous force. HMPE ropes have additional benefits as well: they float, they are resistant to wear, they slide smoothly through the eyelets of ships, and they do not break down in the presence of salt water and sunlight. Pound for pound, they're eight times stronger than steel.

HMPE was first developed in the 1960s, but fibers spun from the plastic weren't commercialized until 1990. The breakthroughs that allowed their creation, new techniques for spinning synthetic thread from gelatinous feedstocks, were the same that gave us Kevlar and other bulletproof fabrics. It's a small thing, these ropes, but one of countless "small" innovations that collectively make every step of this supply chain that much safer, efficient, and economical.

Ropes affixed, the *Netherlands* makes a turn to port, still traveling at a brisk six to eight knots. Back at the harbor pilot station, I can see the ship out the window as it approaches. It turns to starboard, now heading 340 degrees on the compass, its speed dropping below six knots. As it nears, it eclipses the lights of the port behind it. Even Beth, who has seen this dance a thousand times before, pauses to look up. "It never stops amazing me when one goes by in the main channel," she says quietly.

Barring exceptional circumstances, like the temporary freeze of manufacturing in Asia early in the pandemic, every year for decades, the number of containers arriving at the Port of Los Angeles grows. The freight flowing through the port has swollen from the equivalent of less than half a million twenty-foot shipping containers in 1981 to more than 9 million in 2019—an annual growth rate of 11 percent.

The number of shipping containers arriving at ports elsewhere in the United States, and especially ports on the East Coast, such as New York and Savannah, is growing even faster, however, which means that as a proportion of the total number of containers coming into the United States every year, the share claimed by the Port of Los Angeles is shrinking.

Despite all this growth, ships don't call at the Port of Los Angeles as often as they once did. This is because the ever-growing flow of containers is carried on fewer, but much larger, ships. The result for harbor pilots is that in the past twenty years, the number of ships they each have to navigate into port has been slashed from about 600 to 800 a year to about half that, says Craig.

As the ship passes us, the channel narrows to just over a thousand feet across. To my untrained ear, the width of the channel sounds ample for a ship 167 feet wide.

But it turns out that a channel that is seven times wider than the beam of the *Netherlands* is a tight squeeze, after all, on account of the enormous volume of water the ship pushes aside as it plows upstream.

The first obstacle it encounters as it glides through the water is the USS *Iowa*. An old battleship launched in 1942, she once carried Franklin D. Roosevelt to a critical meeting in Tehran with Winston Churchill and Joseph Stalin. Now she's a floating museum moored at Berth 87, dead in the water and a nuisance for all large ships who pass her. As the *Netherlands* approaches the *Iowa*, John orders the helm to lower the ship's thrust to "dead slow," bringing her speed down to about two knots, the same as a leisurely walk.

Every time a big ship approaches the *Iowa*, the bow waves press the *Iowa*'s rusting hulk against massive bumpers that keep the old battleship from scraping against the seawall. After a large vessel passes her, the rush of millions of gallons of seawater into the canyon carved into the water behind that vessel sucks the *Iowa* away from shore, forcing her, briefly, to strain against her mooring ropes.

This situation is made worse by the slight difference between the depth of the channel and the maximum draft of a big ship, which can be as little as a few feet. The totality of all these forces is that putting a ship as colossal as the *Netherlands* through a channel of these dimensions is more like forcing a fencepost down a rain gutter than floating a paper boat down a stream.

Just ahead is the Vincent Thomas Bridge. After passing the *Iowa*, John stops the ship's engine completely. The rest of his maneuvers, until the moment the ship is alongside the berth, he'll complete using only the vessel's residual momentum, like an airline pilot coming in on an emergency dead-stick landing.

It's low tide, so as the *Netherlands* comes under the bridge, the highest point on its superstructure comes within ten feet of the bridge's underside. If this were high tide, clearance would be just eight inches. Regulations say that a ship must be at least two and a half feet lower than the bottom of the bridge, which is just enough to account for variability in the height of the bridge as it sags in summer heat or under the weight of heavy traffic, and also to account for the subtle flexing up and down of a ship at its midpoint, known as hogging and sagging.

Immediately after clearing the bridge, John must turn the entire ship, its engines still at a standstill, 45 degrees to port. At the same time, he must bring its bulk to rest in a berth that begins 330 feet from the Vincent Thomas Bridge, a distance less than a third the length of the ship itself. What he's about to do is the equivalent of a stunt driver parallel parking a car in a spot that's just long enough for it, after coming in at high speed, throwing over the wheel, and skidding sideways to within an inch of the curb, tires smoking.

To make the turn, John radios the tugs on his port side to begin pulling, gently. At the same time, he's using the bulk of the ship itself, the enormous resistance of water pressing against its broad, flat side, to slow the ship as it makes its turn. Even so, the ship will drift—a process called transfer—and continue in the direction it was already heading. The entire thing is like skidding, slowly, on ice, only messing it up means the horrible crunch of hundreds of thousands of tons of metal as the ship misses its mark and collides with either the far side of the basin or one of the ships moored just ahead of where it's supposed to dock.

As if all that weren't enough, during this turn, as with any turn made by any vessel, the point on its hull around which it's turning, known as its pivot point, is constantly moving either fore or aft. In a car, you turn the wheel, and the car turns about a set point, but in a ship, this is never the case, and the pilot's mental model of its behavior must always include this.

Even after the ship's forward momentum has been arrested and she's roughly parallel to the berth where she's supposed to tie up, the high-stakes game of getting her precisely and safely maneuvered isn't over. What was a game of feet becomes a game of inches, with forces no less mind-boggling than the ones that have governed the travel of the vessel across thousands of miles of ocean.

A ship roughly four American football fields long, or three and a half soccer pitches, and 167 feet wide that extends below the surface of the water to the depth of a four-and-a-half-story basement is displacing a volume of seawater equivalent to 105 Olympic-size swimming pools. Using hugely powerful tugboats and bow thrusters, the harbor pilot must move a 1,200-foot-long steel wall against the dock, which means squeezing all the water between those sheer cliff faces of concrete and steel out from between the ship and pier.

To make matters more complicated, the forces acting on the ship become nonlinear as it comes to within a few feet of the dock. At first, there's a huge amount of pressure pushing back against the ship as it's pushing all that water aside, but then, says Craig, it's as if "the bubble bursts."

"At some point, that water runs out from between the ship and the berth, and if you're pushing too hard, all of a sudden you're flying toward that dock," he adds.

In this context, "flying" means any speed faster than about a third of a mile an hour, a speed fast enough, given the forces involved, to obliterate the pier-side bumpers and do millions of dollars' worth of damage to ship and pier alike.

An additional challenge during these especially tricky final minutes of the ship's journey is that with all the containers stacked up on her deck, Captain John can't even see where the pier is. Everything must be done with a "docking mode" on his iPad, and he must trust that he placed and calibrated his satellite navigation antennas perfectly on the ship. Using both GPS and GLONASS, its Russian equivalent, he can get centimeter-precise navigation data.

In addition to moving the ship into the dock at 0.1 knots or fewer, he also has to make sure that the front and back of the ship are perfectly parallel to the pier, because if either end comes in even a tiny bit faster than the other, it puts tremendous, hull-crunching force on that point of the ship.

From the bridge, issuing terse commands to the crew on deck, and via radio, to the three tug captains, John is in total command of the largest mobile, man-made object on Earth. When we drive, our minds expand to make the car a part of our body. That extended embodiment, the fusion of person and object, is a trick our ancestors must have evolved soon after they mastered the use of wooden spears and stone axes. It works no matter the size of the tool at hand.

The instant the ship touches the fenders that keep it from making direct contact with the dock, John orders the tugs to "push full" in order to keep the ship pinned against the dock. If he doesn't, the ship will just bounce right off, as obedient of Newton's laws as any other physical object, no matter its size.

As soon as the ship is pinned, seamen on board the ship throw ropes to shore, each with a monkey's fist knot balled at its end. This knot is heavy and dense enough to allow a rope to be flung a great distance. Indeed, it's so effective at adding a great deal of weight to the end of a rope that a variant known as a slungshot was popular among nineteenth-century street gangs in the United States and was so lethal that it was eventually outlawed. Two years before he became president, Abraham Lincoln defended a man accused of using one to murder someone.

The longshoremen on the pier, of a special type known as linesmen, retrieve the monkey's fists and pull on the attached ropes, which at the other end are secured to the massive mooring ropes, of the same kind used to attach the ship to its tugs, each as big around as a person's thigh. Mooring ropes are so heavy they are often dragged to the proper bollard—the steel nub on a pier that looks like a gigantic mushroom—by a truck. Then a team wrestles the loop at the end of the mooring rope around the bollard.

Only now can the lines be made fast, using hydraulic winches aboard the ship. A ship this size requires sixteen mooring lines, a spider's web of lines pulling on the ship fore and aft, as well as directly toward the pier.

At last, harbor pilot John's job is over. He returns control of the vessel to its captain, disembarks down a gangway, and is whisked away in a car back to the pilot station. Once there, he may have an hour or two to rest, or no time at all, in which case he's straight back onto a pilot boat to do it all over again.

In the pilot station, on a whiteboard above the plush chairs where harbor pilots sit awaiting their next assignment, there's a reproduction of an illustration from an old children's book. Above an idealized 1940s family, mother in pearls, father in a suit, children keeping their elbows off a white tablecloth, is the mantra:

OUR EVERYDAY LIVING
DEPENDS ON
HARBOR COMMERCE

Chapter 6

LONGSHOREMEN
AGAINST THE MACHINE

With the ship fast against the pier, the container with our USB charger inside has finally made it to America. Now the frenzy of activity required to ingest it, and thousands of other containers, into America's consumer market can begin.

On the quay, a symphony of coordinated activity springs to life. Everything is a race against time in a system designed to extract order from inherent chaos. One tugboat captain has compared bringing a ship into port to throwing a wedding, so complicated are all the interlocking dependencies required to get thousands of containers off of a ship in the proper order, and as many back onto it. Every second is dollars incinerated, congestion is the norm rather than the exception, and making money—or as often, not losing it—depends on keeping a ship quayside for as short a period as possible. Before containerized shipping, it could take weeks to unload a ship. Today, it takes as little as twenty-four hours.

At the Port of Los Angeles, there are seven major container terminals, but one of them is special. It represents the future. The future, as ever, is lots and lots of robots.

At 220 acres, the TraPac container terminal is about average for a container terminal in this port. That's about the same area as the portion of New York's Central Park between where it starts at 59th Street to just above the Metropolitan Museum of Art, at 86th Street. A small portion of its area is given over to repair, maintenance, and offices, and a significant portion of it, about half, is only used at times of highest demand.

The rest of the TraPac terminal is a bewilderingly dense thicket of shipping containers made possible only by the complete absence of humans from their midst. Ten thousand containers per acre can be stored in the fully automated portion of TraPac. It's by far the highest density of any container terminal in the United States, and easily among the highest in the world. While there are now others, TraPac is a pioneer in the U.S.: the first fully automated port in the country.

On a blindingly sunny L.A. morning, I'm squinting at expanses of airport-white concrete presided over by the largest robots I've ever seen in my life. One of them is, by dimensions at least, literally the largest robot on Earth. I can't help but be impressed by how smoothly they all move, gliding from one spot to another, on tracks or rubber tires, gentle giants grazing on containers as big as school buses.

Marc Mendez, head of IT for the port and my guide for the morning, explains how containers exactly like the one carrying the USB charger we've been following since Vietnam are unloaded, sorted, and passed on to trucks headed for California's Inland Empire and beyond.

In the chronology of our charger, it's U.S. port arrival day plus one. That is, March 1. In two days, the Centers for Disease Control and Prevention, having botched the rollout of its own test for coronavirus and clamped down on ad hoc testing, will lift its restrictions on all such tests. Along with our charger, hundreds of shipping containers full of things routinely made in Vietnam that will be hoarded or panic-bought in the coming months, from personal protective equipment to web cameras and furniture for home offices, are also being unloaded.

The normally robust flow of goods through U.S. supply chains is about to be thrown into overdrive.

Step one for all of these goods at the TraPac terminal: a ship-to-shore crane plucks a container off of the moored ship. This is one step even the most advanced ports have yet to automate, although the more sophisticated ones have remote-controlled cranes. There are just too many variables to account for, and too much at stake. Ships vary in their dimensions, and the twistlocks that secure shipping containers to one another are surprisingly diverse for something so simple. Some are automatic, but many must be released by hand, which means humans scrambling over stacks of containers and doing one of the most dangerous jobs at the port, working alongside cranes and containers that could crush them in an instant.

Containers above deck are also secured with lashings, and each must be undone by hand by strong men with purpose-built crowbars.

Shipping is full of these odd juxtapositions of relatively puny humans and enormous objects designed to contend with equally enormous forces: from the person at the helm turning a wheel a foot in diameter to deflect a rudder two stories tall, to the sailors tossing ropes to one another to begin the mooring process for a vessel that weighs a million times as much as they do, to longshoremen scampering up, down, and across stacks of containers the size of whole neighborhoods just to release the next box to be grabbed by the office building–size crane hovering above.

The overwhelming majority of ship-to-shore cranes require that longshoremen sit in cabs suspended directly above the gripper portion of the crane, which means staring between their feet for hours, looking straight down through a plexiglass floor in a box shuttling back and forth, out over the ship, back to the pier, each time accelerating smoothly from a dead stop to a fast run. The number of "moves" per hour achieved by a good crane operator makes them essential to the productivity of any port, and they're compensated accordingly.

Half of all longshoremen make salaries of at least six figures, for which they have their unusually long-lived and powerful union, the West Coast's International Longshore and Warehouse Union (ILWU), to thank. Special skills, like crane operation, garner even higher wages.

Cranes grab shipping containers by their corner posts, the same points at which they're secured to one another when on a ship. A shipping container is really a strong rectangular metal frame with relatively weak corrugated sheets of steel in between. They can't be picked up or pushed just anywhere—only, like a car being jacked up by its frame in order to replace a tire, at specific points. Once deposited on the shore, the crane releases the corner posts of the container, and as quickly as the operator can move the trolley in which he rides, it shuttles back over the ship while simultaneously raising its gripper to clear the stack of containers on board.

In most ports, a ship-to-shore crane deposits a shipping container on a yard tractor, which is a sort of funny-looking semitruck. These trucks have only half a cab, just wide enough for one person, so as to increase visibility. Container-bearing trailers are then towed elsewhere in the port, where other sorts of cranes, smaller than the ship-to-shore behemoths, relieve them of their containers and move them into temporary storage.

At TraPac, there's none of that. What happens after containers are delivered to the dock by ship-to-shore cranes is all completely automated. In its peculiar combination of technologies, it is unique not only in the United States but in the world. That has led to unprecedented efficiencies for a U.S. container terminal and also, in the months just before I toured the port in October 2019, a conflagration in local politics.

▶ ▶ ▶

In March 2008, the ILWU signed a contract that sealed their fate—and they've been fighting it ever since. In exchange for an increase in pension benefits to more than $95,000 a year, the union granted the companies that own and operate the terminals at the Port of Los Angeles—shipping giants like Maersk, MSC, and Cosco—unlimited rights to automate these ports, including all manner of computer-controlled and even fully automated cranes.

Such automation was hardly unprecedented—the Port of Rotterdam pioneered modern port automation beginning in 1993—but at the time it was unheard-of. Even today, only 3 percent of the world's shipping terminals are automated. The other 97 percent rely on humans to control everything from ship-to-shore cranes to the trucks and stacking cranes that move and organize containers once they're off a ship.

Like all technologies, port automation is becoming cheaper, easier to roll out, and more effective with time. It's still hugely expensive, and it can cost hundreds of millions of dollars to retrofit an existing terminal. But the upside is machines that work twenty-four hours a day, the ability to stack containers more densely and efficiently, and operational costs that decline with time rather than increase, as wages tend to do.

Another big advantage of port automation is that it doesn't go on strike. Strikes happen irregularly at ports on the U.S. West Coast, but because these ports are choke points through which so much cargo flows, strikes can have an enormous impact. In 2002, a shutdown of twenty-nine West Coast ports so snarled the movement of goods through the combined ports of Los Angeles and Long Beach that President George W. Bush interceded to force the ports to reopen; he then threatened the union with reclassification in a way that would make future strikes impossible.

The tensions between longshoremen and their employers go back a century or more. In 1934, a strike by West Coast port workers lasted three months. Since the 1960s, all major port strikes have been about one thing: the introduction of technology that makes ports more productive but, temporarily and sometimes permanently, reduces their need for workers.

In 1960, the ILWU and the organization representing employers of union members, the Pacific Maritime Association, signed the first Mechanization and Modernization Agreement. It paved the way for containerized shipping, arguably the single most impactful technology in the history of globalization. As containerization spread and its impact accelerated, longshoremen went on strike again in 1971, this time for 130 days, shutting down all fifty-six ports on the West Coast.

Every time the union and the Pacific Maritime Association have to negotiate a new contract, it's the same story: technology has moved on, and someone is going to either lose their job as a result or will have to be retrained for a different role. The year 2008 was no different: the ILWU allowed for more automation in the future, which turned out to be, as ever, a Faustian bargain. Beginning in 2020, the results at the Port of Los Angeles would be the elimination of entire classes of jobs, up to 500 a day at a single terminal, mostly of the longshoremen who drive the tractor trucks that move containers inside ports.

One result of the 2008 contract was the nearly fully automated TraPac terminal I found myself in, a kind of ghost city full of gangly moving contraptions and, in the yard at least, absolutely no people. Another result was the equally automated Long Beach Container Terminal's Middle Harbor facility, which uses different sorts of technology but operates on the same principles. (There are only so many ways to efficiently run a port.)

By the end of 2020, fully a third of the capacity at the combined ports of Los Angeles and Long Beach was automated.

When I arrived in the fall of 2019, the port itself, and the government of the city of Los Angeles, had just been convulsed by the ILWU's last-ditch attempts to curtail a prime example of the automation they'd agreed to in 2008. In the summer of 2019, there were packed public hearings, attempts to block construction permits, intercession by L.A. mayor Eric Garcetti, and a war of words between the ILWU and APM Terminals, a subsidiary of Maersk, the largest container shipping company in the world.

The resulting compromise meant that at the largest terminal in the port of L.A., APM would bring in a total of 130 automated straddle carriers, or autostrads, which are thirty-foot-tall rolling cranes that pick up shipping containers and move and stack them within ports. These autostrads would eventually replace nearly all of the truck drivers within the port itself. In exchange, APM agreed to retrain 450 union mechanics in maintenance of the autostrads and to help "reskill" 450 more union members for other tasks.

Throughout the fight, APM maintained that its 2008 contract with the union allowed it to automate as it saw fit and that it would move forward with bringing in the autostrads no matter what the union said or the city itself decided. This didn't stop the L.A. City

Council from, at one point, withholding authorization for APM to install charging equipment essential to making its all-electric straddle carriers functional.

There are a great many paradoxes inherent in automation of ports and, indeed, in any portion of a supply chain, and this particular battle at the port of L.A. is worth lingering over because it so perfectly embodies all of them at once.

The first paradox of automation is that it can actually increase the bargaining power of some portion of the workers within an industry. The ILWU is famous—in some circles, infamous—for extracting extraordinarily high wages from employers for what is mostly blue-collar work. In part, this is because automation has the overall effect of making labor an ever-smaller portion of the expense of running a port, even if individual workers are seeing their wages increase. This can happen when the total number of workers shrinks, and it can happen even when the total number of workers grows, as long as the operating expenses of all the machinery around them grows faster.

As automation takes over many tasks and makes some workers redundant, it also makes the employees that remain that much more productive. These workers, many of whom have specialized skills and knowledge even if their work is technically blue-collar, have the ability to shut down that much more economic activity should they choose to walk off the job. At ports, this is especially true, since they are, as has been amply demonstrated through the power of striking transportation workers all over the world, choke points in the global supply chain.

Thus, for workers whose skills are especially in demand, think of coders in Silicon Valley, or sufficiently well organized, like unionized dockworkers, automation can make them more powerful, not less. The more than 7,000 full ILWU members who work at the Port of Los Angeles have translated that power into salaries in excess of $100,000 a year, free health insurance, and full pensions. (Union representatives often point out that this level of compensation doesn't reflect the pay of "casuals," a pool of hourly workers of equal size, who are only called up when extra hands are needed, don't get benefits, and start at $25 an hour.)

Counter to the narratives that usually come to mind when we think of automation—Luddites smashing looms that robbed them of a living wage—it can and does lead to better wages for some. On the other hand, along with globalization, in the past fifty years it has also led to polarization of wealth and wages and a hollowing-out of the middle class.

As I wandered in and around the Port of Los Angeles, I was surprised to discover people who worked in the facility or were otherwise connected to it who were unsympathetic to the aims of the longshoremen's union. None of them would speak on the record, but there was, for some, a general attitude that those who were part of the union were so much better off than everyone else with a blue-collar job in the area, in terms of compensation, that they didn't deserve much sympathy. This could be read as either Stockholm syndrome for the American system of capitalism, in which we're accustomed to working-class jobs being stressful, dangerous, and poorly compensated; garden-variety envy; or some mix of the two.

But at least one group of workers who feels their interests are in opposition to those of the longshoremen aren't at all shy about expressing their views. Drivers who pick up containers from the port and take them to warehouses, typically within a hundred miles or so, vent on private Facebook groups about members of the ILWU.

"Hopefully automation kicks in and then all these lazy . . . longshoremen start collecting unemployment," one wrote. The job of these drivers, known as drayage, is brutal in its economic reality. They're paid by the load, not the hour, and many work for small trucking companies or are owner-operators themselves.

Many of these drivers are excited about automation, says Lidia Yan, CEO and cofounder of NEXT Trucking, a technology platform that provides an app on which drivers can pick up gigs, because it means they spend less time waiting at individual terminals to pick up containers.

Turnaround times—the minutes between when a drayage truck enters the gates of a port and when it leaves—bear this out. When I visited the TraPac terminal, it boasted the lowest turnaround times in the Port of Los Angeles, as little as forty-five minutes, when the average for Southern California ports is ninety minutes or more.

There are also environmental reasons why ports are embracing automation. The combined port of Los Angeles and Long Beach remains the biggest single-point source of air pollution in Southern California, despite enormously costly efforts to bring its emissions under control. These efforts include new international laws that force ships to use cleaner-burning fuel when approaching port. What's more, at L.A./Long Beach, ships must plug into the local electrical grid when dockside, rather than running their enormous engines twenty-four hours a day just to power their electrical systems.

But in the port itself, all of the human-driven yard trucks and many of the container-sorting cranes, oversize forklifts, and the like are all still burning diesel fuel. When a port makes the switch to automated systems, it often means a switch to all-electric systems as well, or at least hybrid electric.

There is an important but not often remarked-upon connection between autonomous and electric systems, worth noting because, while the two will have enormous impacts on the future of transportation and automation, when combined, their impact has the potential to be even greater.

As any autonomous systems engineer will readily explain, electric drives—that is, electric motors—and their attendant control systems are easier to automate than mechanical ones, which are less responsive and precise. This slippage between command and response can be overcome, as it is in autonomous long-haul trucking operations, but it matters a great deal more when you're trying to precisely move thirty-ton shipping containers with centimeter precision.

Port operators must also consider the enormous expense of replacing all the shorebound infrastructure in a port. At the Port of Los Angeles, operators ultimately have no choice: a decades-long campaign by the city to make its port greener has culminated in a 2030 goal of having all cargo-handling equipment be zero-emissions. But since automation technology is now relatively mature, any port operator investing billions of dollars on a scale of decades would be crazy not to upgrade their facilities by electrifying them and automating them at the same time.

Thus, the efforts of the city of Los Angeles and its port to become greener have, in ways that are publicly acknowledged by port operators, been yet another driver of automation.

Less than a year after I visited the port, shipping volumes went through rapid swings as factories in Asia shuttered, demand for shipping to the United States cratered, then factories reopened and demand surged, at first to make up for shipments missed in the early months of the pandemic. A year after the pandemic first struck, ports on both coasts were hitting new records as Americans shifted their spending from services to goods. In lieu of going to the theater, Americans bought TVs; instead of going to the gym, they purchased home exercise gear; and to keep their kids and themselves occupied, they ordered mountains of toys, crafts, and electronic devices—nearly all of which came from Asia.

The ensuing global Covid-19 recession might seem like an interruption to all this automation, but it is just as likely to accelerate it. Research by Mark Muro and associates at the Brookings Institution has found that it's recessions, and not the good times, that makes automation happen more quickly.

This might seem counterintuitive—why automate when millions of laid-off workers make it easier to find employees?—but it happens because during recessions, firms' revenue and profits fall much faster than wages do. This mismatch incentivizes companies to adopt whatever cost-saving automation they can in order to protect their bottom lines.

In addition, many companies are realizing that in an age of pandemics, robots have a unique advantage—they don't get sick.

Chapter 7

THE LARGEST ROBOTS ON EARTH

Here's how a port as automated at TraPac, which in one way or another represents the future of all shipping, actually works.

Ship-to-shore cranes at TraPac are still driven by humans who are physically embedded in them, even though remote control is possible and has been implemented at a handful of the world's most automated ports. The longshoremen in TraPac's cranes are real-life examples of the science fiction staple of pilots in giant mechanized suits, except that instead of battling city-destroying Kaiju, they rise each morning to cooperate with equally tremendous seaborne giants transporting consumer goods from one continent to another.

These cranes are bottlenecks in a port, and it is absolutely essential they operate at maximum capacity. The number of "moves" a crane can make per hour is closely watched and made public, part of the way terminals market themselves and tout their efficiency. In order to make them operate as quickly as possible, at TraPac cranes drop containers directly onto the concrete of the pier itself, rather than onto a waiting truck, as at most of the world's ports.

Inside one of those containers is our USB charger. Up to this point in its journey, it's been in the hands of humans who directed the machines moving it. But now, there's a break with many important consequences. This is the point at which the machines take over. From here, the port becomes the sole domain of Marc Mendez, head of IT at this terminal.

Automated straddle carriers in a festive shade of vermilion, each with four spindly legs ending in pairs of wheels, roll up to the containers just deposited on the pier by the ship-to-shore cranes. These autostrads, as they're known, are similar to the devices that were the subject of so much agita elsewhere in the Port of Los Angeles, at the APM terminal. TraPac's autostrads are made by Kalmar Global, a Finnish company.

The twenty-eight autostrads at TraPac are, somehow, whimsical. They accelerate quickly but move slowly, topping out at a speed barely above a brisk walk. They're tall and gangly, like marionettes. They pause to allow others of their kind to cross in front of them, even when by human logic the autostrad already in motion must surely have the right-of-way. They are three stories tall but, all-electric and rolling on rubber tires, almost completely silent. In their group behavior—in the way they literally swarm the containers deposited for them on the quay—they seem like ants or, more accurately, like robots imbued with ant-like logic, a discipline so well elaborated in academia and industry that it has its own subfield, known as swarm robotics.

Like worker ants carrying a dismembered beetle back to the mound, autostrads carry containers from the ship-to-shore cranes to one of a handful of other locations. Some eventually make their way to what may be the largest robot on Earth, which happens to be an automated, rail-mounted gantry crane designed to lift containers from the autostrad worker ants and hoist them atop a railroad car.

Gantry cranes are simple enough devices. A gantry is a post-and-lintel arrangement of a pair of legs with a span of metal in between them, as ancient a design as Stonehenge. Adding a crane to a gantry, one that can slide across the width of its central beam, and putting the entire apparatus on rails that allow it to move perpendicular to the motion of the crane gives you an object of tremendous stability and potentially enormous size, and which is able to lift pretty much anything.

TraPac's railroad gantry crane looks like both an enormous, angular aircraft in flight and a signature building by Rem Koolhaas. The largest of these sorts of cranes has a "wingspan" of 164 feet. TraPac's gantry crane can pick up shipping containers from autostrads that pull alongside and drop them onto flat railroad cars strung into long cargo trains, on parallel rail lines, four trains abreast. This means one such gantry crane can build up to four trains of containers a day, each of them miles long.

A key ingredient of that level of productivity is that these cranes are nearly fully automated. They are capable of operating without human intervention; but just in case, there is a human "in the loop," verifying that the way is clear at a couple of critical moments in the operation of the cranes. For the most part, however, it's the automated system that must clean-and-jerk the container up into the air, over its own support structures, and onto a waiting railroad car.

Rail is important enough for the combined ports of Los Angeles and Long Beach that a public/private partnership spent $2.4 billion to build a belowground rail line, called the Alameda Corridor, in order to allow freight trains to pass through Los Angeles to the two ports without impeding vehicular traffic. But, as is the case in most ports in the world, the majority of containers that move into and out of the Port of Los Angeles are carried on the backs of trucks.

However they will ultimately leave the port, most containers are carried into the container yard, the place where they idle between being taken off a ship and being put on some sort of intermodal transport, like a train or truck. (The Port of Hamburg is exploring using a Hyperloop, the ultrafast transport system proposed by Elon Musk, as a form of intermodal transport, but such a system would be hugely expensive and take decades to complete.)

Not far from where the autostrads picked up their containers, they place them on the concrete a second time. Here, they're picked up by a different sort of gantry crane, known as an automated stacking crane. These cranes exist solely to stack containers in orderly arrays, like building-scale Lego blocks built up into a child's notion of a city skyline. The results are all jagged roof lines and "buildings" snug against one another. These stacks stretch as tall as five containers high and as wide as eight across, with rows of containers stretching hundreds of feet inland, in great conurbations known as blocks.

In between each block, which are the size of those long, anonymous apartment buildings you see in sunbelt exurbs, are narrow lanes, just wide enough for the legs of the gantry cranes to pass. These blocks and these automated cranes allow twice the density of containers to be stored in a given area at a typical port, a TraPac spokeswoman tells me. And while this is not the most automated port in the world, it's the only one with this particular setup of robot autostrads handing off containers to automated stacking cranes. Elsewhere, the straddle carriers are driven by humans, or there are autostrads doing all of the porting and storing of containers.

The entire reason for the existence of container yards and the stacking cranes that rule them, in most of the world's hundred or so large, globally significant container ports, is that it would be horribly inefficient for ships to disgorge containers directly onto the trucks or trains that will take them inland. When a ship carrying the equivalent of 20,000 units of containerized shipping disgorges half its cargo, that's 5,000 forty-foot containers, because the unit of measurement in container shipping is the less-often-used twenty-foot container. If the average truck with a flatbed semitrailer attached is a little over fifty feet long, the length of the line of trucks that would have to queue up to accommodate all those containers would be fifty miles long, bumper to bumper, or about a hundred miles long to allow for a truck-length gap between each of them. Now imagine trying to get all 5,000 of those trucks to queue in exactly the right order. And then multiply it by multiple ships a week, with many of them sitting quayside simultaneously.

The solution to this problem is the same one we find in countless other branches of what's known as operations research, and it has unexpected connections to everything from the architecture of computer networks to the flow of patients through emergency rooms.

In a port, the solution to this problem is those big, lumbering automated stacking cranes building city-block-long, apartment-building-size stacks of shipping containers. Those endless rows of steel boxes are, if you think about it, a buffer between container ships and the trucks and trains they must hand off their goods to. Intermodality at this scale wouldn't work otherwise.

Having a big buffer of containers allows the quayside cranes—the scarcest resource in any port—to operate flat out whenever a ship arrives, pulling containers off as quickly as humanly possible. Then, with inhuman efficiency, the ant-like autostrads rolling around on their octuplet of wheels grab the containers and shuttle them to the automated stacking cranes.

In these areas of the port, automation, and the algorithmic decision-making that drives it, is of the utmost importance.

For the optimal operation of this buffer in the middle of the port, the goal is simple: the order in which containers are stacked in the yard should reflect the order in which they are going to be put onto the human-driven trucks that will take them out of the port. When shippers' trucks come calling, the container needed next should, as often as possible, be on top of the stack on which it's sitting and as close to the end of the row as possible.

Because if it's not—if the next container a truck driver has showed up to pick up on that day is buried at the bottom of a stack, under four other containers—it means unnecessary delay as the automated stacking crane has to dig for the container by pulling off the four on top and dropping them somewhere else, where they might also cause future delays by burying some other container.

It's easy to see how a problem like this could quickly get out of hand. At the worst-performing ports, fully half of the moves made by automated stacking cranes are "unproductive."

The tricky thing about this problem is that there is no perfect mathematical solution. Part of the reason is that there's just too much randomness in the system—ships and trucks showing up early or being delayed, equipment failure, and human error in the few places that humans are still allowed to do anything. But, as with so many other intractably complex problems, there are *better* solutions, and they're extremely valuable. A better solution can shrink the footprint of a port while making it more efficient, saving millions of dollars a year in operational costs.

In 1908, Agner Krarup Erlang, a Danish mathematician, was hired to solve a similar problem for the Copenhagen Telephone Company (CTC). Like all telephone companies, CTC wanted to know how many circuits it needed to build to assure that a given number of people in a town or city could expect to connect when they needed to. In this problem, Erlang, a mathematical prodigy from a young age, encountered the perfect foil for his facility with statistics and probability, and his solution would go on to change the world.

Erlang first proved that if you know the average number of calls per day made by a group of people, you could predict with a reasonable level of accuracy the call volume at any given time. This allowed Erlang to express, in precise mathematical terms, how often, given a certain number of circuits in a town's telephone network, a person might be disappointed by their inability to make a call. Given these numbers, accountants at the telephone company could thus establish how many circuits the company needed to build and at what point building any more would mean extra expense with insufficient additional return in terms of increased availability.

By 1920, Erlang had solved the same problem for the number of telephone operators needed to handle a given volume of calls, since at the time telephone exchanges were conducted by humans, almost always women, who would connect calls directly by physically plugging one circuit into another on a telephone switchboard. Later, the fundamental mathematical unit of the capacity of telephone switching systems he invented would be named for him; to this day, it is called the Erlang.

Erlang's work formed the foundation of queuing theory, which is literally the study of waiting in lines, and specifically the study of the mathematical models of how lines operate. (There's a separate but related discipline—the psychology of queuing—that is applicable to pretty much all of us in the form of that anxiety we experience over whether or not we've chosen the right line at the grocery store.)

Queueing might seem like the product of our modern, built world, but it shows up in countless disciplines—any time a resource is constrained but you need it when you need it. The ubiquity of queuing theory illustrates something much deeper: queuing is a basic biological activity. The need to buffer resources is why our bodies store fat. It's also why the camel has a hump and the humpback whale a rich store of blubber that nearly drove it to extinction once humans figured out it could be converted into another form of useful stored energy: whale oil.

Queuing theory is everywhere, and the pressing need to solve the problems it describes has driven mathematicians to invent ever more sophisticated tools, all of which were made feasible by the advent of electronic computers. Many mathematical solutions to the seemingly simple problem of how to stack all those shipping containers use "genetic algorithms" inspired by Charles Darwin's theories of evolution; others build on those algorithms with artificial neural networks and other forms of machine learning. Like every other problem of routing and resource allocation, the problem of a port is uniquely amenable to the kinds of artificial intelligence that characterize our age. Or, perhaps, having such tools in hand, we have reconstructed our ports so that they can be managed by them.

Once a container is in a stack, it may be moved about in order to make other containers available or moved to the end of the block that is closest to where drayage trucks will receive it and finally take it out of the port proper. This process is called grooming, and, once again, it's uniquely enabled by the automated nature of these stacking cranes. Automated stacking cranes never rest and don't require overtime pay for a third shift, so they can spend their nights, and any other time they aren't ingesting containers from autostrads or dropping them on trucks, grooming their block of containers. It's as if they have an obsessive-compulsive need to ceaselessly rearrange their charges, these multiton steel bricks, up to the limit of other predefined constraints, like the trade-off between using more energy to do so and whatever increased efficiency more container moves might offer.

The average "dwell" time in a port is just a few days, which means that within a week, the average port ingests and expels the majority of containers in its yard, tens of thousands of them. Making this happen requires a feat of coordination at the gates of a port no less fantastical than the ones required to get containers off a ship, only this one involves thousands of humans instead of just a handful of crane operators.

Near the end of my tour of the TraPac terminal, I'm standing amid the lanes that semitrailer trucks back into in order to receive their designated shipping containers. Of all the places in the port, this is the one most entangled within a skein of automation, mathematics, human labor, and the complicated history of how that labor organizes itself to protect, or fail to protect, against the brutal vagaries of unfettered markets.

There are also lasers, an actual web of them, shooting across the lanes, bouncing off of little mushroom-shaped caps so that they hit trucks and containers, allowing the systems that have partially automated this area of the port to know precisely where those critical bits of kit actually are, down to the centimeter.

This is the one place in the port where humans can be near the gigantic autostrads, sliding to and fro, at one moment looming and the next out of view, behind stacks of containers. Drivers are given a prearranged time to arrive, scheduled by an algorithm so as to minimize the time they must wait in the port while also maximizing the efficiency of the sequence of containers lifted by the crane that day. Drivers back their trucks in and then get out and stand in what look like phone booths positioned on little concrete islands in between each lane. Here, as in factories filled with industrial robots, the key to keeping soft, eminently crushable, water balloon–like humans safe is to segregate them from the machines completely.

Once a driver is in the safety kiosk, the invisible lasers shooting through the air allow the system to identify where the twistlocks are on the flatbed trailer and precisely lower a container onto those locks. Once they have the all clear, drivers hop back into the cabs of their vehicles and pull away.

The same procedure happens in reverse for trucks dropping off containers. For maximum efficiency across the entire system, it's best if trucks are dropping off and picking up containers simultaneously, but it isn't always possible. There are so many different shipping companies and shippers, all with competing priorities, that a driver might find themselves dropping a container off at one terminal and then taking their empty trailer to a second terminal to pick up a different shipping container.

Inside one portion of the office buildings that sit near the entrance to the terminal, a TraPac spokeswoman introduces me to a longshoreman. His job could not be more removed from his forebears', who, like chefs with their knives, each carried their own custom-made, wood-handled, wicked-looking metal hooks in order to grip and sling crates, barrels, and sacks full of goods. These men worked outdoors year-round, deep in the sweltering or freezing holds of ships. This longshoreman, by contrast, is sitting in an ergonomic office chair at an expansive desk in a quiet, open-air office. Before him is a bank of monitors, each showing a feed from a different camera perched on the automated stacking crane as it lowers a shipping container onto a semitrailer truck.

These systems are mostly automated, but like practically every automated system in industry today—whether it's the one driving a truck or flying a drone—a human is "in the loop" for "exception handling," that is, any situation the automated system can't handle. A box of half-eaten sushi and a coffee mug are the only decorations on his desk. He talks to me as he deftly maneuvers the crane with a small black joystick, and he seems self-conscious about me being a journalist and about the presence of my PR handler, who is technically part of management, or at least not part of his union. The whole situation, the irony of his presence in the belly of a beast designed to expunge him and as many of his kind as possible, is not lost on either of us.

I ask him a few questions as he works. The container he deftly hoists without interrupting our conversation is identical to the one that carries our gadget—identical, in fact, to all shipping containers.

He says the job is fine, easier than it used to be. Other terminals are less automated, he continues, and are more efficient as a result, in his opinion. My handler stands by, radiating a certain nervousness. A few moments of silence pass between us. Then he offers, wryly, "You don't need a college degree to do this."

Chapter 8

THE LITTLE-KNOWN, RARELY UNDERSTOOD ORGANIZING PRINCIPLE OF MODERN WORK

It's been said that nothing in biology makes sense except in the light of evolution, and the same could be said of the relationship between the ideas of Frederick Winslow Taylor and the character of modern work. Yet the average person, and even many experts in the fields of manufacturing, supply chains, economics, and labor, don't know his name, nor the name of his discipline, scientific management. Even those who do generally have only the vaguest understanding of his significance. For most pundits and historians, he is at most a caricature, a straw man, either the patron saint of the economic miracles of efficiency and mass production or the devil who sped up and routinized modern life, robbing workers of both agency and rest.

Frederick Taylor's impact has been compared to that of Charles Darwin, Sigmund Freud, and Karl Marx. No less a personage than future supreme court justice Louis Brandeis made Taylor's ideas famous in 1910 by employing them in a pivotal case argued before the Interstate Commerce Commission. In 1914, Lenin wrote that scientific management was how capitalism extracted the most from its beleaguered subjects; in 1918, he reversed himself and said scientific management would be essential to building a functional Soviet state. Scholars of Taylor have argued that the French, who adopted his ideas even faster than industrialists in his home country, could not have held the Kaiser at bay in World War I without him. Even one of his harshest critics, management theorist Peter Drucker, credited Taylor and scientific management with winning the Second World War.

His ideas were debated in the halls of Congress in 1912 and then banned from use in U.S. government armories. Workers subjected to management informed by his techniques went on strike dozens of times in the span of thirty years, their pamphlets filled with cartoons and essays inveighing against him by name.

There are many reasons for the erasure of Taylor from the history most of us learn in school. The first is that events at the end of his life, and his own sloppiness and even cruelty in implementing his ideas, have forever stained his legacy. Almost no one who works in his shadow or is heir to his ideas—and that's just about every manager and management consultant in every firm of any sophistication anywhere on Earth, at this point—wants to admit any connection to him. The second reason is that as influential as Taylor's ideas were, explicit references to his work by his disciples faded in the middle decades of the twentieth century. Some scholars attribute this to the shift of manufacturing in the United States to factories overseas. As we exported our productive capacity, we forgot those who once enabled it.

The most important reason Taylor has been forgotten is that, for better and for worse, his insights, philosophies, and methods have become the water we swim in. Now that we are all cogs in the elaborately interconnected, increasingly mechanized, ever more production-obsessed machine we call the global economy, ideas that seemed novel in his time—he did most of his work in the late nineteenth century and disseminated his ideas in the first decade of the twentieth century—have been so completely internalized by us all, no matter our role or field, that they now seem obvious.

The ideas Taylor espoused, and as with all prophets, the ones he came to represent once he was no longer able to speak for himself, are by now inseparable from the way business is done. Like an invisible puppet master, his ideas animate the way capitalism, in its alchemical capacity to turn the base metal of goods and services into money, works.

Taylor's ideas were so powerful that they quickly spilled over, beyond the walls of the factory and shop floor where he formulated them, into other areas of human enterprise, from transportation and logistics to medicine and the then-burgeoning fields of knowledge work. From there, his adherents applied them to the conduct of everyday life. The bestselling 1948 book *Cheaper by the Dozen*, later made into a movie by the same name, depicts the real-life family headed by Lillian and Frank Gilbreth, who were Taylor disciples, management consultants, and efficiency experts. (At the time, those three things were basically synonymous.) The Gilbreths applied Taylor's work and their own, primarily in motion studies, to the raising of twelve children.

In *Cheaper by the Dozen*, the memoir of their childhoods, Frank Gilbreth Jr. and Ernestine Gilbreth Carey wrote of their father, "He'd walk into a factory like the Zeiss works in Germany or the Pierce Arrow plant in this country and announce that he could speed up production by one-fourth. He'd do it, too."

Elsewhere, they wrote, "Our house at Montclair, New Jersey, was a sort of school for scientific management and the elimination of wasted motions—or 'motion study,' as Dad and Mother named it. Dad took moving pictures of us children washing dishes, so that he could figure out how we could reduce our motions and thus hurry through the task."

If you've ever tracked your pace, counted your steps, monitored your sleep, called something a "life hack," or consumed an article, book, or podcast about enhancing personal productivity, you've used and internalized methods pioneered by Taylor and his disciples.

"At night, each child had to weigh himself, plot the figure on a graph, and initial the process charts again after he had done his homework, washed his hands and face, and brushed his teeth," wrote Frank and Ernestine. "Mother wanted to have a place on the charts for saying prayers, but Dad said as far as he was concerned prayers were voluntary."

You might be wondering what any of this history has to do with getting goods from the factory to your front door. The answer is that while Taylorism once applied primarily to the factory floor, something fundamental has shifted in how we live and work. As materials and finished goods move ever more quickly and efficiently through the global supply chain, and information has come to be transmitted more or less instantaneously and for free, the walls of the factory have dissolved. Every day, more and more of what we do, how we consume, even how we think, has become part of the factory system.

This is not a metaphor. Functionally, our world has become a single, increasingly well-integrated factory, with the methods of mass production embedded not only in how we make things but also in how we distribute and consume them, whether for work or leisure. In our drive to extract maximum value from every moment of our lives, Taylor's methods and philosophy are how we live now.

The elements of production that once had to be brought together in a single place are now strung out along supply chains tens or even hundreds of thousands of miles long, as goods crisscross the globe before becoming finished products that are delivered to our homes. By extension, we are all now living in one sort of factory town or another. If this sounds strange or far-fetched, it could be because most of us in "developed" countries are so far removed from the manufacturing and distribution of goods. People who work in service or knowledge work tend not to be connected to the messy, often gritty world of actually making things, concentrated as it is in the sprawling factory systems of the developing world and (still) in parts of the United States they never have cause to visit.

During the pandemic, the list of who was considered "essential" made it apparent who is most integral to the global factory system. Those who made our goods and those who delivered them were both essential; those who worked in hospitality, travel, and many other parts of the service economy were not.

No matter how much an economy "dematerializes" itself, making and exporting software and intellectual property instead of steel and fabric and cars, dumping money and workers into health care, finance, marketing, and an ever-expanding and ever more abstract assortment of knowledge work, we still have to eat and house and clothe ourselves. Not to mention all the keeping up with the joneses–type consumption that we pile atop the essentials, what economists call "status competition."

The factory was invented because producing goods at scale necessitated getting a bunch of people together under one roof. Its earliest examples were the textile factories of England, where workers, almost exclusively women who had once spun wool at home, were gathered with those who wove it into cloth. Together they attended to machines that helped the owners of these factories wipe out what had been employment at home—the original gig economy—for what one scholar estimates were two out of three women in England.

Later, this kind of factory, and the attendant push toward total vertical integration, reached its apotheosis at Henry Ford's massive River Rouge Complex, where raw materials such as rubber and iron entered at one end and finished Model Ts came out the other.

Today's factories are nothing like that. But when we think of them, the image many of us still have is of raw materials going in one end and finished goods coming out the other, with smokestacks and machines and workers on assembly lines in between. Modern factories, even gigantic ones of the kind in which tens of thousands of workers assemble iPhones or Toyota Corollas or Nike tennis shoes, are but "final assembly" plants—the last step in a long supply chain of countless suppliers and intermediates who make or modify materials and parts before they arrive at this final staging area.

A typical supply chain—that is, a typical distributed factory system—might start with natural gas harvested in the panhandle of Texas. Then it's piped overland to a plant in Houston where it's chilled to a temperature of −265°F. In this state it's loaded onto a liquid natural gas tanker and transported more than 18,000 miles, across the Atlantic, through the Suez Canal, around the southern tip of India, and through the Malacca Strait, to China. There, it's piped into another plant and, through the miracle of synthetic chemistry, turned into plastic. Then it's transported by rail to a second location in China where it's spun into polyester thread, which is trucked to a third place in China to be woven into polyester fabric. Then it's shipped to Cambodia to be sewn into garments for a recognizable Western sportswear brand. Only *then* does it become a part of the supply chain covered in this book—from the factory to our front door—which in its own way has become part of the distributed factory system that resulted in a finished product in the first place.

Supply chains for processed foodstuffs can be just as convoluted, and those for more complex goods such as consumer electronics or automobiles are a hundred times as complicated, drawing on sprawling and ever-shifting supply chains for a seemingly endless array of individual materials and components.

All of the manufacturing facilities and the supply chains that feed them and distribute their products are Taylorized to one extent or another. Measuring the output per minute, hour, and day of every machine, worker, and system in these factories and warehouses is so bog-standard as to barely warrant mention. Yet the level of attention to productivity in any of these facilities would have shocked workers and managers of the first century of the Industrial Revolution, when something as basic as accurate timekeeping was an uncommon luxury.

Taylorism's success in all of these places has meant its penetration into every step along the journey of a made thing, as well as lateral transmission into pretty much every other kind of work. But to understand any of that, we have to start with the man himself.

Chapter 9

HOW A MANAGEMENT PHILOSOPHY BECAME OUR WAY OF LIFE

Frederick Winslow Taylor was born in 1856 in the Germantown neighborhood of Philadelphia, the only child of a Princeton-educated lawyer and an ardent abolitionist. He was raised with all the privilege a boy of such pedigree could expect—a stately house on a hill, tutoring from his devoted mother, an education that included eighteen months of travel and study in Europe. He got into Harvard and had plans to be a lawyer, but for reasons that remain obscure, he decided on a very different career.

America was in the throes of a crippling economic depression when he leveraged family connections to become an apprentice pattern-maker and machinist at Enterprise Hydraulic Works in Philadelphia.

To be born and raised in Philadelphia at that time was a bit like being born and raised in Silicon Valley in the late twentieth century, or Shenzhen, its equivalent in China, today. With its relative wealth, proximity to Appalachian coal, and high concentration of machine shops and other manufactories, Philadelphia was a hotbed of cutting-edge technologies imported from elsewhere and quickly improved upon by American innovation. English engineers were brought in to help American industrialists copy their successes in the United

Kingdom, bringing the coal and steam era of the Industrial Revolution to the shores of the new world.

Fred Taylor entered the pump factory at a time of tremendous change in how things, and especially machines, were made. These innovations would have a profound effect on the amount of wealth that could be generated in the production of things, raising the standard of living of millions of Americans to a level never before seen anywhere in the world.

So distinct was this time of both innovation and rising wages for skilled workers that it constitutes its own phase of the Industrial Revolution, says Robert Allen, a professor of economic history at New York University Abu Dhabi. Unlike our present era, in which every one of us is more productive than ever but the wages of the middle class have been more or less stagnant since the late 1970s, this was a time of both rapid invention and rising fortunes, a period of optimism but also upheaval, as America recovered from the Civil War and absorbed wave after wave of immigrants.

Taylor, who was slight of build and self-conscious about the fancy way he talked, took pains in the machine shop to make himself into one of the guys. As recounted by Robert Kanigel in *The One Best Way*, the authoritative—and only—contemporary biography of Taylor, he peppered his speech with curse words to make himself seem coarser than he was. He told no one where he lived, which was at home with his parents in the fancy part of town.

A hard worker and a quick study, he joined an informal guild of skilled metalworkers whose prized abilities gave them a great deal of power over how their work was accomplished and at what pace. Theirs was the age of craft production, as opposed to the later systems of mass production. It was a method of manufacture that would be recognizable to medieval stonecutters, cobblers, blacksmiths, farriers, coopers, and woodworkers. Every item was made by a skilled craftsman. (They were almost all men, outside of the textile trades.) To create a complicated pump or other device, these craftsmen would start with the first part of that device and subsequently machine, by hand, with chisel and file and lathe, all subsequent parts so that they fit the first one perfectly.

Later, owing to some combination of his ability and family connections, Taylor became a "gang boss" in charge of a number of machinists far more skilled and senior than himself. It was at this point that Taylor's easy demeanor changed. Having been one, Taylor knew that the workers under him could work far more efficiently than they did. He would later admit that they were not, for the most part, lazy so much as the victims of perverse incentives. At the time, almost all productive work of any kind was compensated by the piece. The problem was that if workers produced more, bosses would eventually cut their rate. The two sides were stuck in a sort of prisoner's dilemma with no ready solution.

Years later, Taylor would cite his time at the pump manufacturing company when he argued that he had a unique understanding of the psychology of skilled workers. A full accounting of Taylor's life renders this argument self-serving at best, and perhaps even delusional.

The workers under Taylor, whom he motivated primarily by following through on his threats to fire underperformers, detested him so completely that they regularly threatened him with violence. His family feared for his safety as he walked home alone along deserted railroad tracks. His ability to raise productivity at the pump factory despite the ill will he caused between management and labor convinced the bosses to keep him around.

Taylor would later helm a failed attempt at building a company around a novel process of papermaking. Eventually, he struck out on his own as one of the world's first management consultants, and then he "retired" in 1901. He wrote two books—*Shop Management*, published in 1903 and intended for experts, and *The Principles of Scientific Management*, published in 1911 and aimed at a lay audience. Both were subsequently translated into every language spoken in the world's industrializing economies, from Europe to Japan, and became foundational for manufacturers the world over, from Renault to Mitsubishi Electric.

Taylor's methods were simple enough: He would observe workers performing a task, break it into its component parts, and record the amount of time it took them to perform each subtask. Workers were offered incentive pay—say, 50 percent higher wages—if they could use what Taylor learned about how the best of them did their jobs to produce at a level that was often two or three times higher than what they had managed before.

It's worth pausing to appreciate that the simple act of recording precisely how long it took a worker to complete a task down to fractions of a second, a necessity for examining the dozens or hundreds of individual steps in their labors, was not only new but almost impossible until just about the time that Taylor started doing it. One of Taylor's first promotions was to the role of time clerk. At the time, "punching a clock" wasn't even a thing, because time-stamping clocks had yet to be invented. The first was patented in 1891. Taylor started at Midvale Steel Works, his second job after working at the pump factory, in 1878. The job of a time clerk like Taylor was to verify that everyone showed up when they were supposed to and that no one left early.

Integrating precision timekeeping with the piecework systems of the day—a process fundamental to nearly all corporate blue-collar jobs today, where productivity per minute, hour, and day is constantly measured as a matter of course—was yet to come. Performing the kind of productivity measurements that would become the hallmark of the efficiency movement and scientific management required the appropriation of a tool previously used almost solely for a very different purpose.

The first mass-produced stopwatches in America numbered only 400 and were produced from 1858 to 1861 by the Waltham Watch Company. Known as "chronodrometers," they were intended for timing horse races—a hugely popular entertainment at the time. Along with clocks and pocket watches produced by the same company, they were among the first widely available timekeeping devices in the world to be made from interchangeable parts. By the 1870s, stopwatches were common at horse races and sporting events, and they were also used in the military to time artillery shots.

But it was Taylor and his disciples who would bring these instruments into the factory. Mass production and interchangeable parts made these devices widely available and also eventually drove universal adoption of the system of scientific management. In later years, scientific management would come to be so closely identified with the stopwatch that men in suits skulking about factory floors, stopwatches and notebooks in hand, were treated with open hostility by some workers.

In his *Principles of Scientific Management*, Taylor sought to turn his ideas into stories that were accessible if not strictly true, in a sort of proto–Malcolm Gladwellian effort to market his ideas. In what he made the founding myth of scientific management, he told a probably apocryphal tale of a laborer named Schmidt, whose job it was to load pig iron into a cart, and whom he supposedly convinced to work three times as hard for about 40 percent more pay. The founding myth of Taylorism and scientific management, and therefore all of management consulting, is a story of a steelworker in a race not unlike the one between John Henry and the steam drill, only in Taylor's telling, instead of the laborer dying at the end of the story, he gets a raise.

The story of Schmidt appears to be based on Taylor's experience at Bethlehem Steel. Here, Taylor arrived at a figure for how much pig iron manual laborers should be loading by asking the best of them to work as hard as they could for an hour. He then deducted 40 percent from the amount they managed to load in this sprint for rest breaks. Multiplying that number by the number of hours in a workday yielded the amount of iron he expected all workers to load. Workers who couldn't meet this pace, which was most of them, were fired. After two years of consulting, Taylor was also fired. He then billed the company $100,000 for his services, or more than $2.5 million in today's money. "Ordering people around, which used to be just a way to get things done, was elevated to a science" by Taylor, wrote historian Jill Lepore.

Taylor was neither the inventor nor the sole practitioner of what would come to be known as Taylorism, a name most often used pejoratively to describe what had previously been known as scientific management. He was just scientific management's richest and most visible hype man, and also apparently quite charming when in the company of other management consultants and the bosses who signed his checks.

Practitioners of the broader discipline of efficiency were a kind of Left Bank intelligentsia of management theory, except instead of Gertrude Stein and Ernest Hemingway discussing writing in a Parisian café, it was Taylor, Louis Brandeis, the Gilbreths, and Henry Gantt (inventor of the Gantt chart, still used by project planners today) discussing the uselessness of wasted motion on the factory floor in an apartment in New York City. It was in that apartment where they settled on the phrase "scientific management" as the name of their philosophy.

Taylor died in 1915, at age fifty-nine, three years after being grilled for three days by a special committee called to investigate a strike at the Watertown Arsenal. Watertown was where the U.S. government manufactured cannons and other armaments, and where Taylor's assistant had instituted a system of scientific management that was universally loathed by workers. Molders at Watertown who were told to pour a gun carriage in less than half the time they usually did wrote to their boss: "This we believe to be the limit of our endurance. It is humiliating to us, who have always tried to give the Government the best that was in us. This method is un-American."

There are a number of reasons Taylor's methods took off soon after the promotion of his ideas by Brandeis and the publication of his last book. Chief among them was that, as historian Daniel Nelson wrote, "he told people what they were ready to hear."

To understand why the world was ready for Taylor's message, it helps to know a bit about the fertile soil into which he dropped the seeds of his ideas. Even before Taylor set out on his own as a management consultant, in 1893, "cost accounting systems, methods for planning and scheduling production and organizing materials, and incentive wage plans were staples of engineering publications and trade journals," wrote Nelson.

Factories were something new in the world, and complicated to boot, a collection of humans and machines of a scale never before seen outside of armies or construction sites. It took decades for those who ran them to learn how to manage the workers within them. The more complicated the products of a factory, the more that management fell to the workers themselves, since no one else knew better how to do their jobs. When Taylor entered the machine shop, a transition was underway in which managers of variable backgrounds were being replaced by engineers granted the responsibilities of management. Engineering itself was just becoming an accredited profession, and as the scientific method was applied to the products of engineering, so too were engineers primed to apply it to the management of the humans who worked with machines.

In an age in which the use of interchangeable parts was not yet universal and the assembly line was just getting started—that is, the age before Ford's system of mass production had taken over the world— it was often the case that every complicated machine was different from every other and had to be constructed by hand. The speed at which people worked was the biggest bottleneck to productivity and revenue. Today, we talk about speeding up a production line and we mean making the machines on it run faster. In Taylor's time, during the turn of the nineteenth century, it meant only one thing: speeding up the humans.

As mass production with machines took off, making humans faster became, paradoxically, even more important. As machines sped up the production processes for which they were responsible—stamping, milling, molding, and the like—humans had to keep pace. Conceived at the dawn of mass production, Taylorism became inseparable from it.

This speedup of machines, and the increased demands it made on the humans that worked alongside them, was the product of a number of other enabling technologies. The first, and the one with which Taylor was most intimately involved, was reliable production of large quantities of steel.

When Taylor entered his first shop, steel still had to be shipped from England. Small amounts of it could be used to make tools hard and durable enough to shape items made of cast iron, on a spinning lathe or sometimes even with a chisel held in one hand, as if the machinist were carving wood. By the time Taylor retired, steel was being made in vast quantities by a variety of then-cutting-edge but increasingly reliable methods, and it went into absolutely everything that was transformative in this era. Locomotives, the tracks they ran on, ships, armaments, engines of every kind, farm equipment, sewing machines, tools, the earliest automation in factories, and the first automobiles all were made possible because of cheap and readily available steel.

Steel's unique properties mean it can be cast into any shape, then machined to incredibly fine tolerances. Steel tools and steel measuring devices in the hands of skilled and obsessive craftsmen could yield truly interchangeable—and durable—parts. Interchangeable parts could then be produced in massive quantities. Eventually, Henry Ford borrowed the idea of the conveyor belt from slaughterhouses, where carcasses were carried from one worker to the next on overhead hooks, and married it to the system of interchangeable parts he had already borrowed from other automakers.

On October 7, 1913, Henry Ford's first assembly line clattered to life in the company's urban Highland Park, Michigan, factory. Asked later whether they were influenced by Taylor's ideas, Ford and his deputies took great pride in insisting they'd never heard of the man. And yet countless commentators have intuitively—but incorrectly—leaped to the conclusion that Taylor's and Ford's ideas about management were directly linked.

The most critical thing to understand about scientific management, Taylorism, the mass production system that came to be known as Fordism, and many other early efforts to measure and enhance the productivity of workers in factory and factory-like settings is that many versions of these ideas sprang up independently and then quickly came to influence one another. Like so many other inventions and discoveries, from the light bulb and the steam engine to calculus and the theory of natural selection, the preconditions necessary for a variety of smart and dedicated people to arrive at a new idea were present in many places at once. To historians of innovation, this phenomenon is so common it has a name—"multiple discovery."

The real power of Taylorism isn't that Taylor and his disciples were directly responsible for all subsequent systematic optimization of human effort. It's that Taylorism came to represent a set of commonsense answers to a widespread question: How can the human side of mass production be made faster and more efficient? That something like scientific management proved to be the answer for many different people who might have at first been only vaguely aware of one another should hardly be a surprise given all the wonders that scientific inquiry was churning out at the time, especially in the latter part of the Victorian era.

Ford's system of production took on a life of its own, in part because Ford was eager to share it. At the turn of the century, in an era before television and even radio, factory tours were viewed as a valuable means of marketing and a way to demonstrate the sophistication and power of a firm. Sears, Roebuck and Co., at the time strictly a mail order company, gave daily tours of its blocks-long facilities in Chicago. A 1906 tour of its mail-sorting operation impressed Ford and shaped his thinking as he worked with his engineers to create a system of mass production.

In the same way, thousands of visitors passed through Ford's gargantuan River Rouge Complex, at ninety-three buildings the largest integrated factory in the world when it was completed in 1928, and many were influenced by the production system they witnessed.

In the meantime, Taylorism underwent rapid evolution as people all over the world applied its basic notions of how to increase efficiency through a combination of pushing workers harder and identifying the best way for each to do their job. A basic tenet of Taylorism was the extraction of knowledge from skilled workers in order to train all workers how to do their jobs in the "one best way." With mass production came automation, and with automation came a new way to take knowledge out of the heads of workers and embody it in a system that didn't need their hard-won abilities.

An early example of this was the metal stamping machine. With cheap and readily available steel at hand, manufacturers realized they could use giant presses to cut and shape sheets of steel into uniform parts that required almost no additional machining. It was a huge departure from the difficult process of casting that preceded it, which required that trained patternmakers, the role in which Frederick

Taylor originally found himself, craft shapes from wood and then hand them off to other skilled craftsmen who would use them to make an impression in sand that could be filled with molten iron.

Economists call the process of replacing experienced craftspeople with machines that can be run by more easily trained workers "deskilling," and it's integral to the modern economy. From preparing a meal in a fast-food joint to driving for Uber, countless careers that used to require a significant amount of experience and skill have now been turned into jobs that require hardly any at all.

From the beginning of the twentieth century to 1915, adherents of Taylorism brought his methods to nearly 200 American businesses. But the most direct and immediate impacts of Taylorism occurred outside the United States. In 1914, the Allies found themselves under increasing pressure to produce material for the First World War, but they had neither a reserve of factory workers nor additional raw materials with which to accomplish it. The solution was efficiency.

"Even before the war, French military officers had recognized the potential of scientific management for arsenal operations and had introduced Taylor's methods in at least one plant," wrote Nelson. "After 1914, as they struggled to increase production, they increasingly relied on scientific management for the manufacture of shells, arms, explosives, motor vehicles, and airplanes."

Managers in France followed the letter of Taylor's philosophy in a way Taylor himself never did. Taylor had, after all, called himself a friend of workers and claimed that his primary method of increasing productivity was to determine how to make a worker's job easier. At a French navy yard, metallurgist León Guillet created a planning department that was responsible for time and motion studies. It also offered additional wages to the most productive workers. Rather than firing underperformers as Taylor might have, Guillet treated his workers with respect and apparently realized the promise of Taylorism at its best—increasing efficiency and productivity without harming those subjected to it.

Others in Taylor's circle adopted similar strategies when bringing scientific management into firms. Frank Gilbreth insisted that both management and labor sign off on any changes to working conditions intended to increase efficiency. After the death of her husband, Lillian Gilbreth paid special attention to the condition of workers and wrote a book called *The Psychology of Management*, subtitled *The Function of the Mind in Determining, Teaching and Installing Methods of Least Waste*.

"[Scientific management] has demonstrated that the emphasis in successful management lies on the *man*, not on the *work*; that efficiency is best secured by placing the emphasis on the man, and modifying the equipment, materials and methods to make the most of the man," she wrote. "It has, further, recognized that the man's mind is a controlling factor in his efficiency."

That squishy "controlling factor," which was the subject of so much of Mrs. Gilbreth's consulting and writing, was the thing that the engineering mind of Fred Taylor seemed unable to grasp, and which so often led to worker revolt against his methods.

Brandeis also recognized the importance of this factor, and he wrote about it in a 1915 essay in *Harper's Weekly* titled "Efficiency by Consent." He argued that under a fully developed system of scientific management, "the greater productivity of labor must not only be attainable, but attainable under conditions consistent with the conservation of health, the enjoyment of work, and the development of the individual."

"In a democracy it is not sufficient to have discovered an industrial truth, or even the whole truth," he continued. "Such truth can rule only when accompanied by the consent of men."

Brandeis estimated it could take up to twenty-five years for a properly managed firm to reach a state in which it was both as efficient as possible and sufficiently humane in its treatment of workers. But what Brandeis missed—what was inconceivable at the time he wrote his essay, during an era of rapidly expanding membership in labor unions—was that efficiency wouldn't always be extracted from workers by consent.

Lillian Gilbreth, meanwhile, found other ways to encourage Taylorism that did not require negotiations with unions or the invitation of management. She brought it into the home. "The search for the One

Best Way of every activity, which is the keynote today in industrial engineering, applies equally well to home-keeping and raising a family," she wrote in 1955.

Gilbreth invented the kitchen island, and also the work triangle layout of modern kitchens, where sink, stove, countertop, and refrigerator are all intended to be within arm's reach of a home cook. She came up with the idea of a trash can that opens with a foot pedal, shelves on the doors of refrigerators, and light switches in walls. You cannot stand in a modern kitchen, or even turn on a light as you walk into a room, without beholding her work.

In consulting for General Electric, she interviewed housewives in order to refine the designs of household appliances. She wrote that if housewives could be more efficient, they could have enough time to work outside the home, as she did. Despite her eagerness to dispense advice to those who did, she did not cook. Speaking of herself and her husband, Frank, who died in 1924, leaving her to raise twelve children alone, she said, "We considered our time too valuable to be devoted to actual labor in the home. We were executives."

She designed a series of kitchens for corporations to use as part of various promotions, and she made claims about them that were eerily similar to the ways that Frederick Taylor asserted his methods had saved time and expense for his corporate customers. One, called the "Gilbreth Kitchen," she designed for the Institute for Women, an organization sponsored by the *New York Herald Tribune*. In a 1931 compendium called *The Better Homes Manual*, a writer described a test of her "labor-saving kitchen":

> [A] cake was first made in a typically haphazard kitchen. We kept a record of every motion and every step taken in this process. Then an exactly similar shortcake was prepared in the Herald Tribune Kitchen, which has the same equipment and utensils as the other kitchen, but has them arranged for efficiency. The results of this test were so startling as to be almost unbelievable. The number of kitchen operations had been cut from 97 to 64. The number of actual steps taken had been reduced from 281 to 45—less than one-sixth!

Gilbreth's move into the home was out of necessity. Despite being an engineer herself, and the uncredited author of Frank Gilbreth's last book, contracts for her services dried up after her husband's death. But her talent for making what were then some of the most time-consuming tasks of everyday life more tractable demonstrated, more than anything else up to that point in time, the broad applicability of scientific management. Frederick Taylor codified and popularized the fruits of the efficiency movement, but it was Lillian Gilbreth who taught the world that it was possible to use careful observation and the scientific method to speed up just about anything.

The irony of Gilbreth's efforts to save housewives time was that as women working in the home became more capable, the standards of "good housekeeping" rose to match their enhanced abilities. Instead of saving America's homemakers time, she unintentionally burdened them, and eventually all of us, with greater demands and more complex tasks. It was in no small part her work that meant "modern technology enabled the American housewife of 1950 to produce singlehandedly what her counterpart of 1850 needed a staff of three or four to produce: a middle class standard of health and cleanliness for herself, her spouse and her children," wrote historian Ruth Schwartz Cowan.

One of many ironies of scientific management is that by the measure of its ability to reduce the total quantity of humanity's labors, it was a complete failure. Taylorism was in the end not an efficiency but a *productivity* movement.

Taylor himself recognized this. When Upton Sinclair, newly famous as the muckraking author of *The Jungle*, wrote a letter to *The American Magazine* critical of how little a worker's wage increased in Taylor's depiction in his *Principles of Scientific Management*, Taylor responded that Sinclair had missed the point entirely. Throughout history, it wasn't the worker who benefited the most from increased efficiency, or even his boss.

"In fact," wrote Taylor, "a glance at industrial history shows that in the end the whole people receive the greater part of the benefit coming from all industrial improvements. In the past hundred years, for example, the greatest factor tending toward increasing the output and thereby the prosperity, of the civilized world has been the introduction of machinery to replace hand labor . . . And this result will follow the introduction of scientific management just as surely as it has the introduction of machinery."

The privilege of productivity over all other concerns shows up again and again throughout the supply chain. This chain and its components are the machine that gratifies consumer desires, and consumer desires expand endlessly as efficiency makes an ever-expanding variety of goods and experiences affordable and available to us.

Taylor's biggest contribution to our world isn't some now-obscure system of speeding up work. It's levels of convenience never before thought possible, and a pace of life even he could not have imagined.

Chapter 10

RIME OF THE
LONG-HAUL TRUCKER

Picture an eighteen-wheeler. Or maybe you call it a semitruck, a big rig, a tractor trailer, a tractor truck, an articulated lorry (if you're from the United Kingdom), a semilorry (Australia), or just a truck. Odds are you see a large cab in front and a long, rectangular, smooth-sided metal box in back. Maybe it's sporting a logo belonging to Amazon, FedEx, UPS, Walmart, or your local grocery chain. Maybe, if you're a particularly keen observer of what traverses our nation's highways, you imagine a box that says J.B. Hunt, XPO Logistics, Swift, Schneider or Werner Enterprises, which are among America's largest independent trucking companies, carrying freight for whoever cares to hire them.

America has 3.5 million truck drivers, and about half, or 1.7 million, drive long-haul trucks, which include the one you're picturing as well as specialized haulers for automobiles, liquids, fuel, and other hazardous materials, in addition to flatbed trailers with no box on top.

But it's the stock-standard trailer on a big rig, the "dry van" as it's known, sealed against weather but not temperature-controlled, that is the most common way to transport freight in the United States. A lightweight box about six times as long as it is tall or wide, it typically has a hardwood floor and a steel frame. Its walls and roof are a skin of aluminum just 0.04 inches thick, or about three times the thickness of a fingernail. The trailer part of a semitrailer is the most minimal and economical design humanity has yet invented for enclosing a unit of freight that must travel on a road while being protected from the elements.

More than 300,000 truck trailers are produced in the United States every year. If you've ever noticed a Wabash, Great Dane, or Utility logo on the back of a truck trailer or its mudflaps, that's the manufacturer of the trailer, not the brand of the truck or trucking company.

Boxes of goods stacked on wooden pallets go into the backs of long-haul truck trailers hundreds of thousands of times a day, every day, in the United States. Some are transferred in the cross-docking warehouses in California's Inland Empire, where goods are removed from shipping containers and transferred to the dry vans of semitrailer trucks. On this day, March 3, 2020, a box on a pallet, put there more than a month ago in Vietnam, is one of those being transferred. Inside is our USB charger.

On the same day, the World Health Organization (WHO) will declare that the global number of confirmed Covid-19 infections has exceeded 90,000, almost all of them in China. The WHO will also note that the virus has reached seventy-two countries. A panel of experts, including Dr. Anthony Fauci, will submit to Congress testimony declaring that "the potential global public health threat posed by this virus is high, but right now, the immediate risk to most Americans is low." Amazon will inform its Seattle-based employees that one of their coworkers has the virus.

Sealed into its dark, theftproof metal tomb, a chamber not so different from the one in which it crossed the ocean, this box won't leave the truck again until it arrives at its end point in Shakopee, Minnesota. That's 1,785 miles, twenty-seven hours' driving time, and two and a half working days from now, on March 6. While it's en route, the Dow will tumble nearly 1,000 points in a single day, cruise ships will begin testing passengers, and the U.S. Environmental Protection Agency will touch off a run on household cleaners by releasing a list of ones that are effective against Covid-19. Our box will experience countless starts and stops, accelerations and decelerations, noisy rumblings over potholed city streets, the low thrumming of the truck's eighteen wheels over smooth highway, and the subtle expansion and contraction of the trailer with each change in the weather.

There used to be some adventure in this journey, or at least Americans had the idea that moving things in big rigs was exciting. The trucker-hijinks classic *Smokey and the Bandit*, starring Burt Reynolds and Sally Field, was the third-highest-grossing film of 1977. (Number one that year was, of course, *Star Wars*.) And it was hardly an outlier; the late 1970s were host to at least a half dozen films romanticizing the life of truckers, from *Convoy* starring Kris Kristofferson and *High-Ballin'* with Peter Fonda to *Every Which Way but Loose* costarring Clint Eastwood and an orangutan (really).

These days, the journey from port to consolidation warehouse to distribution center, where pallets of goods are sorted onto trucks destined for big-box retailers like Walmart and Target or fulfillment centers, where individual items are stored and sorted into packages for delivery to our front doors—as at Amazon and other e-commerce companies—has become a flat-out race against a digital clock. Drivers must meet deadlines that are as likely to be set by an algorithm and an online bidding system as a trucking company dispatcher or an account handler at a freight-forwarding company.

On its road trip, our box will leave San Bernardino on Interstate 15 and cruise through Barstow before skirting the southern tip of Death Valley. Throughout its journey, it doesn't so much go through towns as run beside them, the interstate having long ago bypassed main streets and downtowns, now relegated to the "business route" forking off from the highway before rejoining it without having captured anything other than local traffic. This doesn't mean a shortage of places to get fuel and stop for a meal or a shower, of course—like rivers and continental shelves, highways create their own ecosystems.

In Las Vegas, our box passes billboards advertising loose slots and washed-up stage acts. Fifty miles later, it winds around the northern tip of Lake Mead, the slowly dying man-made lake that is the primary source of water for 25 million people in Arizona and Southern California. For a moment it touches one corner of Arizona, then enters Utah near the stunning red rock cliffs of St. George. Accelerating to eighty miles an hour and more on long, flat stretches of highway, it passes under big skies and a bright sun. Just outside the ghost town of Sulphurdale, named for the rich ore that was the source of both its boom times and desertion, it hangs a right onto I-70. So swift and efficient is the machine carrying it that barely seven hours have passed. At this point, by federal law, it's about time for the driver of the truck pulling our trailer to take a thirty-minute break.

Our driver is a long-haul, or "over-the-road," truck driver, a person who used to be among the best-compensated blue-collar workers in America. That all ended within a couple of years after President Jimmy Carter signed the Motor Carrier Act of 1980, the culmination of efforts to deregulate America's trucking industry (and transportation in general) begun by Richard Nixon.

At the time, President Carter justified his support of the act as a righteous attack on two things then besieging the American consumer: inflation and the cost of fuel. Deregulation of trucking, he predicted, would lower the cost of goods for Americans by $8 billion a year, $25 billion in today's dollars, while helping America's trucking fleet conserve fuel. He also predicted truckers themselves would "benefit from increased job opportunities."

Carter wasn't wrong. Truck transportation costs dropped 25 percent for a full truckload by 1982 and even more for some shippers of less-than-full truckloads. Truckers themselves started hundreds of new trucking companies, and the new trucks they put on the road were more fuel efficient than the older models they replaced. But truckers who were already in the industry paid a price: increased competition destroyed what had been strong labor unions of long-haul truckers and slashed their wages.

The average trucker in the United States made $38,618 a year in 1980, or $120,000 in 2020 dollars. In 2019, the average trucker made about $45,000 a year—a 63 percent decrease in forty years. What had been a job earning the average truck driver the equivalent of a six-figure income has today become one where starting wages are barely enough to keep new drivers on the road.

The result is what the American Trucking Associations (ATA), the largest trade organization in trucking, frequently calls a "shortage" of truckers. In 2005, according to the ATA, the United States was short 20,000 truckers, but that ballooned to more than 60,000 truckers by 2018, and it could reach 160,000 by 2028, implying something like paralysis for an $800-billion-a-year industry that currently moves more than 70 percent of America's freight, with dire consequences for us all.

Aside from the large trucking companies and truck manufacturers the ATA represents, no one else I interviewed believes there is actually a shortage of truckers. A 2019 paper that counts among its authors an economist from the U.S. Bureau of Labor Statistics declared that while the labor market for truckers is "tight"—in other words, it's challenging to hire truckers because there isn't a surplus of them—it's hardly broken. Thus, concluded the authors, the term "shortage" does not apply.

Raising wages and improving the working conditions of truckers could entice more into trucking and keep more people in the industry longer. Those who have worked in the industry, as well as smaller trade organizations like OOIDA—the Owner-Operator Independent Drivers Association—argue that rather than a driver shortage, which implies challenges in recruitment, trucking has a burnout and retention problem.

It certainly seems telling that the typical large trucking fleet must replace on average 100 percent of its drivers in a year, even as the ATA argues that most of that turnover is drivers moving from one trucking company to another.

As a measure of how many people abandon trucking altogether, there are 3.5 million truckers in the United States, and 10 million Americans have a commercial driver's license, the special certification needed to drive a large commercial vehicle, says Steve Viscelli, a sociologist who studies the trucking industry. "We've had way more than enough workers cycle through this industry [to end any driver shortage]," he adds.

As part of his research for a PhD, Steve actually became a long-haul trucker. He has also interviewed hundreds of truckers, in every segment of the industry and under practically every possible employment arrangement, from the most fortunate drivers, who are salaried employees, to newbies just starting out who are, like most truckers, paid by the mile and treated accordingly. "We've had multiples of this shortage number who got into the industry, got screwed over, lived out of a truck for six months, and lived at basically minimum wage. And then they said, 'Forget it, I haven't seen my family in six months and I made $45,000 on an annual basis, but it's not worth it. I can get a minimum wage job or two or three and at least be home at night.'"

To become a truck driver almost always requires spending months at a community college or a carrier's own driving school, incurring thousands of dollars of debt. Despite all that effort, and additional debt few who choose this path can afford, one in three new truck drivers abandons their employer in the first ninety days.

Talk of a trucker shortage and relatively high wages for experienced truck drivers do seem to help the industry continue to recruit the tens of thousands of drivers it needs to ingest every year in order to replace those it loses. Drivers for private fleets in the United States— think Walmart-employed drivers pulling Walmart-branded trailers— make on average $68,000 a year. But these plum jobs are available only to the most experienced drivers with the cleanest driving records. And just because you see a trailer on the highway bearing a recognizable logo doesn't mean it's being pulled by one of these drivers. Internally, Walmart uses the large carrier Swift, which employs many relatively inexperienced drivers, to haul freight between its warehouses. Amazon does the same, tapping whatever trucks and drivers are available at rates it dictates.

Our pallet of boxes, including the one holding our USB charger, is riding on a truck driven by a person who, as often as not, feels trapped in their job—a "throwaway" person, as one trucker told the *New York Times*. How cynical they are about their industry depends on their age, level of experience, and the personality of their employer. But even drivers at some of the best carriers—surveys indicate drivers are happier at smaller ones—are, by many measures, suffering.

Working sixty to eighty hours a week for less than half the wages your father could have earned in the same trade doesn't help. But the root of this angst isn't merely economic.

As much as any other node in the global supply chain, truckers are subject to the disempowering and demeaning effects of the combination of automation, surveillance, and intensification of work that is frequently a consequence of modern information technology. The converging technologies and economic forces that inspired Frederick Taylor to invent scientific management find pure and too often sinister expression on America's highways and in the cabs of her trucks. In industries where competition is intense and workers interchangeable, the result—stressful working conditions that push people to their limit—is more or less inevitable under current U.S. labor laws.

You see this again and again in industries as varied as trucking, call centers, and fast-food outlets. Wherever it's relatively easy for new companies to enter an industry, the people doing the labor required can be trained in just a few weeks, and individual businesses compete on cost instead of quality, companies will subject employees to conditions that lead to high turnover.

In the clinical language of economists, making a job so awful that your employees quickly burn out and quit is a natural "cost-minimizing response." For all that bosses and management consultants bemoan the high cost of turnover, it's in the end cheaper than improving wages and working conditions enough to eliminate it. For many industries, this means a sort of equilibrium of misery—a natural set point for how difficult, dangerous, monotonous, stressful, and psychologically damaging the average job can be.

▶ ▶ ▶

Long-haul trucking is much like shipping or warehouse work or drayage—essential, invisible, and neglected in our thinking and policies. The plight of truckers is in many ways the same as that of sailors, and the words of Rose George, author of *Ninety Percent of Everything*— "Don't assume the principles of fair trade govern the men who fetch [things] to you"—apply here as well.

Those men include Robert Gallard, age fifty-one, a resident of Orlando, Florida. Like all long-haul drivers, he rarely sleeps at home. On the day I meet Robert, he isn't running the route our USB charger is taking, from L.A. to Middle America, although in his twelve years of driving a big rig, he's run it and just about every other major corridor in the United States.

The journey Robert has agreed to take me on runs up the Atlantic coast, but other than the weather it is not so different from the journey our box is taking from L.A. to Minnesota. Highways, like the vast and trackless ocean, are, for reasons of safety and convenience, all remarkably alike.

On a cold January day, I climb into the cab of his Freightliner truck in the parking lot of a lawn mower factory in Maryland. For the next nine hours, he will talk almost nonstop, even as he braves ice, snow, winding roads, and seemingly suicidal drivers of passenger cars. He will unspool his life story and much of what he has learned in his many grueling years as a trucker, and in the passenger seat I will absorb his strange tales and stranger theories, rapt.

In some ways, Robert is typical of America's over-the-road truck drivers. He's male (as of 2018, 93.4 percent of America's truck drivers are men) and middle-aged (the average age of an over-the-road truck driver is forty-six). In other ways, he's less typical. Around 66 percent of America's truckers are white, but Robert is Black, as are 14 percent of truckers.

Despite being a minority, Robert feels he's rarely been discriminated against by other drivers, at least in ways he can perceive. That doesn't mean he hasn't experienced chronic and systematic discrimination of another kind, stemming from his status as an ex-felon.

Robert grew up hard in North Philadelphia. Born in 1968 and raised in the projects, he turned seventeen the same year the word "crack" appeared in the *New York Times* for the first time. Soon after, while on parole for a different offense, he was convicted of manslaughter and illegal possession of a pistol. He had, by his own account, used it to defend himself when attacked on his way home. By that point in his life he'd already been stabbed once and shot twice.

He would serve twelve years. In prison, he reset his life. In a weird way, he says, it saved him.

After he got out, he bounced around between odd jobs—the sort that felons are limited to, often off the books—before teaching himself all he needed to pass the written portion of the test required for a commercial driver's license (CDL). A CDL is what you need to drive a Class 8 truck, which is what the U.S. Department of Transportation calls semitrailers and other large vehicles such as dump trucks. Robert then went to a community college to pass the driving portion of his CDL exam.

Trucking, despite its reputation, is a skilled trade, and its challenges are varied. For one, truckers must learn how to drive a vehicle of surprising complexity and exacting demands. The majority of trucks still have manual transmissions, and experienced drivers prefer them for the additional control over the vehicle they allow.

More and more trucks have automatic transmissions, part of a push to make trucking more accessible to people with a wider variety of backgrounds and skills, but Robert hates them. Automatic transmissions are, by his account, unwieldy at low speeds and incompetent at highway speeds. They're liable to make the truck buck like a bronco when slowly backing a trailer into a loading dock, smashing bumpers and equipment alike. They're also unable to make the most of the power of a truck's engine when a driver attempts to pass on an incline at highway speeds, leaving a truck either stuck in slow traffic or boxing in other vehicles.

A typical long-haul truck has at least twelve gears, and shifting between them requires not merely matching the truck's gear to the speed of the vehicle, as one does with a stick shift in a car, but also revving the truck's engine to exactly the right rpm in order to match that gear. It's this infamous "double clutch" of big rigs and the mental gymnastics of backing a fifty-three-foot trailer with movable rear axles into spaces where inches count that are the tasks most likely to thwart those who attempt to get a commercial truck driver's license.

On top of the basic mechanics of driving a truck, there are the challenges of doing so safely for an average of 100,000 miles a year. A single crash can mean the end of a career, not to mention the death of those in the passenger vehicle that was probably also involved. The maximum load a big rig is legally allowed to carry is 80,000 pounds, and shippers often do their best to approach that maximum in order to make the most of every trip.

Consider what happens when a passenger vehicle cuts off a tractor trailer on the highway, which according to the average trucker, and my own observations during the 400-mile journey with Robert, happens at least once an hour.

If all traffic is moving at the same speed, darting into the hundred or more feet truckers prefer to maintain between themselves and the vehicle ahead of them might lead to nothing more than an uptick in the driver's stress level. But suppose an aggressive driver of a passenger vehicle is weaving around slower vehicles. Or a car moves in front of a truck and immediately slows down, which happens a lot in heavy in traffic.

It takes 200 feet for a fully loaded tractor trailer to stop when it's traveling fifty-five miles an hour. It takes significantly more distance—a football field or more—for it to come to a stop when it's traveling faster and the roads are bad.

Truckers can't just stomp on their brakes when things get hairy. If they try to stop too quickly, the momentum of their trailer will carry it forward even as the cab of their truck slows, causing the trailer to swing around one side of the truck, a disaster called jackknifing. This typically sends the truck onto its side, and rollovers frequently injure or kill drivers. Plus, trucks on their sides block multiple lanes, potentially leading to a chain reaction of other crashes and fatalities, a lethal pileup that can cost millions. Even if the truck doesn't jackknife, a sufficiently heavy load of the right dimensions can punch through the front of a trailer and into the cab, killing the driver. The stakes multiply if the driver is carrying an excessively heavy, hazardous, or oversize load.

All of this is on my mind as Robert pulls onto the highway, making full use of the on-ramp to accelerate at what feels like a painfully slow pace, despite the surprisingly loud rumble of the truck's engine. From the driver's seat, he radiates a tightly coiled energy, his gaze darting from rearview mirror to blind-spot mirror to the road directly ahead to the road far ahead to the mirrors on the other side of the truck and back, over and over again. Sometimes his eyes narrow when he notices a potentially tricky vehicle, traffic pattern, or change in the condition of the road.

Doing this for an hour, maybe a few hours, maybe the length of a normal workday is one thing, but Robert must do this for as close to eleven hours of legally allowed active driving time a day as he can manage. That doesn't count the additional three hours he might spend waiting around for lumpers to load and unload his truck at any warehouses, factories, or other facilities he visits throughout his day, nor additional time he may, semilegally, mark in his logs as personal driving time to and from a place to sleep.

Sitting in a big rig is to see the world from ten feet up, and it grants a weird sense of power, even omniscience. I'm able to see over the tops of even the tallest vehicles, farther down the road than I've ever experienced in a passenger vehicle.

It's also fucking terrifying. The truck's engine roars, the transmission yowls when we miss a gear, and Robert makes noises of disgust as cars pass us in ways that endanger their drivers' lives and our own. I'm acutely aware of the fact that fatigue kills and that excessive sleepiness is the physiological equivalent of drunk driving. Think of the kind of fatigue that could set in if you've been sleeping in the cab of a long-haul truck for weeks or even months on end, every night a new rest stop or truck stop, cops banging on the window at 2:00 a.m. if you've been forced to pull over on the side of the road because no other accommodation is available, or sex workers doing the same if you've pulled into some stops.

Safety is also impacted by lower wages. Most truckers are paid by the mile, which means they have every incentive to push themselves as hard as they can for as many hours as they can get away with. Research by Michael Belzer, an economist at Wayne State University, found that at a per-mile rate of 60 cents or more, truckers will ease off and deliberately get more rest. Average per-mile rates in 2018 were closer to 35 cents a mile, and some truckers I interviewed spoke of rates as low as 20 cents a mile—break-even territory.

The consequences of low per-mile wages for truck drivers are enormous. In 2017, the latest year for which data are available, heavy trucks were involved in more than 4,400 fatal crashes, up 23 percent from 2010. In the overwhelming majority of those crashes, it was whoever was occupying the passenger vehicle who died. That same year, an additional 107,000 crashes resulted in injury.

Every two years, more Americans die in crashes with heavy trucks than have been killed in combat in Iraq and Afghanistan since the beginning of those conflicts.

It starts to snow. After an hour, it's almost whiteout conditions. Robert becomes, somehow, even more tense. Our route, from the town of Frederick, Maryland, to the even smaller town of Derby, New York, just outside Buffalo, includes some rural highways, the kind designated by three-digit numbers, as opposed to major interstates, which are always two digits. "I guarantee it's slippery on them little state roads," says Robert, toggling between route options on his GPS.

Chapter 11

100 PERCENT OF EVERYTHING

If you're at home, in an office, or at a coffee shop, pretty much everything in your line of sight right now came to you by truck. The array of technology that made its trip possible is considerable.

"There's this idea that trucking is really conservative about technology, but it's not," says Steve Viscelli, the sociologist turned trucker turned academic who has made it his life's work to study the industry.

This reputation might come from the fact that trucks themselves haven't changed much since the 1970s. Almost all semitrucks have 400- to 600-horsepower diesel engines, enough to move forty tons up and down the steepest grade they're likely to encounter. Where makers of passenger automobiles, buses, and medium-size delivery and utility trucks have rapidly iterated from gasoline engines to hybrids and now fully electric vehicles, the physics of long-haul trucking have kept manufacturers using the same big engines that became popular in trucks sent to the front during World War II. Batteries won't hold enough energy to take a long-haul truck any reasonable distance, and the regenerative braking that makes hybrids so fuel efficient means little to a vehicle that spends most of its time on the road cruising at highway speeds.

Inside a truck, it's a different story. Gone are the road maps of yesterday, and while there's still a CB radio, it's hardly the dominant mode of communication it once was.

In Robert's truck, he has two smartphones, one personal and one for work; a laptop on which he communicates with his boss and shippers; an Xbox One and a PlayStation 4 for blowing off steam at the end of his long days; a trucker-grade Garmin GPS; a Canon printer/ scanner for the paper documents he has to both produce and digitize for shippers and his dispatcher; a tablet he uses as a dashboard camera, a backup dashboard camera because in an accident you can never have too much evidence, and a third camera provided by his employer; a federally mandated electronic logging device with a tablet-based interface, which is connected directly to the innards of his truck and tracks his hours on the road; an E-Z Pass and a weigh station pass; a CB radio; and a rat's nest of chargers and charging cables.

Many of these devices are but end points for cloud-based services, from email to route-planning software and apps designed specifically to help truckers find places to fuel up, eat, and sleep. Ever enterprising, truckers have even turned Google Maps into a sort of Yelp for pickup and drop-off spots. Punch any factory or warehouse into Google Maps and scroll down to the reviews, and you'll find they're not from customers but truckers. Here they leave notes for each other about how to handle the staff at any given location, whether they're rude, if there's a break room, where the nearest legal parking is, and which are the best roads to take to reach a facility from the highway.

None of this has changed the fundamentals of the life of a long-haul trucker. A driver with as much experience as Robert should be eligible for one of the better routes coveted by drivers, the kinds that come with salary and benefits and are basically regular loops between two locations—say, a coastal warehouse receiving goods from shipping containers and an inland one where goods are sorted for delivery to stores or regional e-commerce distribution warehouses.

But Robert's felony conviction makes him ineligible for such routes. In this way, he has more in common with drivers new to the trade. Robert is on the road for weeks, sometimes months at a time, lucky to get a few days out of every month at home. Truckers without regular routes must string together one load after another, because they don't make money pulling empty trailers.

The strain of professional long-haul trucking is phenomenal and so relentless that it transforms the bodies of the people living it. A 2018 study of truckers found that 85 percent were on the road twenty-one or more days a month, 70 percent worked more than eleven hours a day, and 77 percent experienced frequent time pressure. Few had a regular schedule, 38 percent reported they rarely or never got a good night's sleep while working, and on average they drove 3,000 miles a week. Seven in ten reported being paid by the mile, and nearly the same proportion reported being stressed about work. Combined with a shortage of breaks, few opportunities for any kind of movement, much less exercise, and the limited food options at truck stops, the result was that 58 percent of truckers had the markers of metabolic syndrome, that cluster of high blood pressure, obesity, and high cholesterol that raises a person's risk for heart attack, stroke, and type 2 diabetes.

Robert is thin for a trucker, almost wiry. He says he gets almost no exercise, but he also barely eats when driving, sometimes going an entire day on a single light meal and a succession of black coffees. They have no effect on him anymore, he adds; drinking coffee is just "something to do."

In the nine hours we're on the road together, Robert stops once, to get $450 worth of fuel, at a TA truck stop in Erie, Pennsylvania. It's clean, well organized, and so like most every other truck stop in America as to be almost indistinguishable from them. Outside, there's just the entrance, a service garage, and a big lot for trucks to park in for the night. When fueling, trucks stop under an awning twice the height of the kind in a gas station for cars. It's bay after bay filled with giant semis gleaming under floodlights that burn twenty-four hours a day. Walking inside the truck stop, I pass under a brightly lit sign that says:

WELCOME
PROFESSIONAL DRIVERS

This makes me feel like an impostor and an interloper, but also like an anthropologist and a spy. No one pays me any mind as I walk about the building taking pictures of things that could not be more mundane and less notable to the dozens of truckers passing through.

In 2016, writer Kyle Chayka coined the term "AirSpace," his name for what Rem Koolhaas warned in the mid-1990s was, in the form of airports and hotels that were already growing architecturally indistinguishable from one another, the dawning of the "generic city." Basically, Chayka's idea is that the reason you can go to a coffee shop in Tokyo, San Francisco, Kiev, or Saigon and they all look exactly the same (blond wood, Edison bulbs, exposed brick) is that social media is everywhere, homogenizing our taste and flattening the world into one indistinguishable miasma of cortados, chill EDM, and Ikea couches.

Truck stops are a funhouse mirror version of AirSpace, where instead of a single tasteful accent color and flourishes of stainless steel, there's an entire aisle devoted to beef jerky; a wall of particolored drinks that are just different combinations of artificial flavor, sugar, and caffeine; and a spinning rack of six-inch folding pocket knives made in China but sporting handles decorated with the American flag, a bald eagle, or, if you're feeling especially patriotic, a bald eagle with American flag wings.

Inside this particular truck stop there's also a country store, and every letter of its logo is jauntily offset vertically from the one before it, suggesting both a colorful southern pronunciation and a deliberate flaunting of the rules of readability, even an unconscious protest against the studied minimalism of the logo of the coffee shops inhabiting Airspace. The country store also has a purely decorative wood-shingled awning that ends at the ceiling, which is one of those pockmarked dropped ceilings that looks like a swarm of beetles was encouraged to bore holes in some other, better kind of ceiling. It's dark outside and the light in here is at the harsh, blue end of the spectrum.

It would be easy for an outsider to look on this scene, especially while contemplating the sameness of all truck stops, and find it dispiriting. But that's a complete misreading of the function of corporate-chain truck stops, the feelings of the truckers who use them, and this

TA in particular, which online gets high marks from truckers. One review from Shanna Lemen, a designated "local guide" on Google Maps: "There was a little Asian lady waitressing in the country pride restaurant that was amazing! Left her a really nice tip for her hard work. Hubby and I pulled off during whiteout conditions and didn't have to wait for a shower . . . that was real nice! . . . Thanks TA."

The sameness of well-run truck stops is a huge relief for drivers whose days are unpredictable in all other aspects, including their duration, the weather, the behavior of other drivers, the orneriness of the lumpers who loaded their truck, and the nature of that day's load. It's the same reason American tourists will visit a McDonald's or a Starbucks, whatever country they find themselves in: there is comfort in the familiar.

For truckers, this sameness is also a question of basic self-care. Many are not able to shower for days, may have run out of clean laundry, and are stuck eating whatever food is available on the road. This TA, with its bank of washers and dryers, salmonella- and norovirus-free buffet, daily specials that include actual vegetables, and familiar layout, is a salve to a weary trucker's embattled soul.

The showers just around the corner from the entrance advertise themselves with a sign that says VOTED BEST SHOWER. It continues:

Fast Wait Times
Just Under 2 Minutes
Average Wait Time
Fresh Towels
High Quality
Towels
Crisp Ventilation
Brisk Air Flow
Clean Vents

All of which suggests showers at other truck stops fail on one or more of these measures.

Back in the truck, Robert tells me another thing he likes about corporate truck stops like this one is that they're monitored better than the dying breed of rough-around-the-edges, mom-and-pop ones. Mostly that means no sex workers banging on the door of his cab in the middle of the night.

Meanwhile, in Utah, the driver hauling our USB charger stops at hour seven of his journey not because he needs fuel or food—he's traveled fewer than 500 miles, using less than half of his onboard supply of fuel—but because as of 2019, the law says he has to. Hours-of-service rules used to be flexible, to put it mildly. It was easy to fake paper logs and accepted across the industry that drivers would lie on their reports, driving longer than the eleven hours they're allowed in a day, or spending more than fourteen hours on the clock total. Often, they accrued those extra hours during long loading or unloading times, when they might nap in their sleeper cabin.

All that started to change in December 2015, when laws mandating use of electronic logging devices came into effect. They were the result of thirty years of political and legal wrangling between insurers, large fleet operators, the Federal Motor Carrier Safety Administration, judges, and independent truckers. The resulting rules were a compromise no one was completely happy with, and the rules continue to be revised as their impact on truckers and safety becomes apparent.

Research indicates the arrival of the logging devices has forced drivers who weren't already using them to comply with regulations about how much and when they can drive, but they've had no effect on the number of crashes. They also appear to have increased unsafe behaviors by some drivers, mostly speeding, as they wrangle with the decrease in flexibility these devices force on them.

A significant amount of freight moved on trucks before the 2015 update mandating logging devices required a full sixteen- or seventeen-hour day of drivers, on account of the amount of time they'd have to spend waiting at warehouses and other facilities for their trailers to be loaded. The long-haul-trucking system was built on the assumption that a driver would drive 400 or 500 miles in a day, says Steve Viscelli, with long load and unload times at either end of that journey. But that's just not possible anymore. Big truck companies haven't been affected because they were already using electronic logging technology, but it's squeezing small- and medium-size operators that were already struggling to gain any edge they could over their larger competitors.

The result is even more strain on drivers than before, says Robert. Again and again, he recounts ways that new regulations or a paucity of infrastructure like rest stops and truck stops squeeze drivers like him. The drivers he encounters are stressed-out and angry, threatening each other over the CB radio when they perceive violations of road etiquette or when another driver misses a gear when climbing a hill and is forced to go slower than he would like, blocking others from passing him.

"[Trucking] is like making a deal with the devil," says Robert, as we pass a giant LED sign showing an image of Donald Trump riding on a tank, the slogan "America Is Back!" in giant letters across the top. "You don't see the family; whatever relationships you have can get strained; there's stress, traffic."

Later, Robert tells me he's fighting with his wife of eight years "constantly."

"I think I'm going to have to divorce her," he says, almost nonchalantly. Being gone as much as he is requires a partner who is uniquely forgiving, and his current marriage to his high school sweetheart, whom he reconnected with after prison, isn't a fit.

Some truckers get around the problem of loneliness and strained relationships by taking their husband or wife with them on the road. In some of these teams, both partners may have commercial driver's licenses, allowing for more lucrative team driving but creating a new kind of strain, as one person must sleep while the other drives. Many truckers bring along pets. Dogs are a favorite, and Robert has his eye on a rescued French bulldog back in Florida.

It's amazing how often truckers I encounter don't have any significant attachments at all—children grown, no spouse, their lives dictated entirely by their need, even desire, to sleep in a different place every night. Trucking remains, or has become, one of those jobs into which people disappear.

By the end of our trip, we're encountering one of the most vexing problems truckers deal with on a daily basis: there aren't enough places in America for them to park their trucks for the night.

America is moving more freight than ever on more trucks than ever, and the number of rest stops and truck stops has not kept up. To make matters worse, many businesses where truckers deliver won't let them stay in their parking lots overnight. Walmart allows individual stores to decide whether or not trucks can park in their lots, which means about half of America's 5,400 Walmarts don't allow it.

By 8:00 p.m., we're close to the end of our journey. Having discovered from reviews on Google Maps written by other truckers that he can't spend the night in the lot of the vendor where we're to drop off our load first thing in the morning, Robert detours to the closest rest stop. It's so packed that trucks are parked along the highway leading up to it, occupying the shoulder of the off-ramp all the way to the rest stop proper, which is completely full. On the way out, trucks are stopped for the night so far down the ramp back onto the highway that the last couple are parked illegally.

"That's the risk you take," says Robert—the risk of being awakened and ticketed by state troopers just for trying to get a night's rest.

The entire time he's driving, Robert's energy never seems to flag. By the end of the night, I'm ready to collapse into a real bed, but he's still going strong, still chatting, still filling the silences with stories and asides, an unceasing chronicle of the ways truck drivers are pinched between ever-tighter regulation of their time and state and local law enforcement who seem eager to give them tickets just for being where everyone who depends on them—that is, all of us—needs them to be.

Chapter 12

HOW "HITLER'S HIGHWAY" BECAME AMERICA'S CIRCULATORY SYSTEM

Just before the electronic logging device of the hypothetical driver hauling our USB charger registers a violation of the rule limiting him to eleven hours of driving time in a single day, he pulls into the Loco Travel Plaza in Fruita, Colorado. Traveling 719 miles while maintaining an average speed of sixty-five miles an hour is a miracle of one of the most important but least appreciated technologies in the entire overland supply chain: roads.

For anything to move across Earth's landmasses quickly, its journey must be transformed into something akin to the journey of a boat on calm seas. Everything that causes friction must be minimized.

That's why the interior of America was first opened up not by rail but by the unbelievable effort required to dig and excavate, with nothing but the physical labor of men and oxen, the 363 miles of the Erie Canal. It took eight years. At the time, barges and ships were the only way to transport large amounts of freight for a reasonable amount of money and in a reasonable amount of time. The physics of transporting goods by water are so compelling that to this day, the

Mississippi River carries a tonnage of freight that dwarfs the total amount carried by any overland route in the United States.

That we move so much freight on our roads—and that technologies only just dawning will move a great deal more in the future, as you'll discover later in this book—is due to many people whom history has mostly forgotten.

Those people include President Franklin D. Roosevelt, civil engineer Thomas Harris MacDonald, and Adolf Hitler. Dwight D. Eisenhower got his name on America's interstate highway system at the end of a decades-long effort to plan and build it because he's the president who finally secured the funding required, not because he had some particular genius in designing it or making it politically palatable. By the time Ike signed the Federal-Aid Highway Act of 1956, drivers had been clamoring for it for decades, with local businesses even going so far as to fund the building of their own primitive highways. Those first "highways" were nothing like what we imagine now.

Well into the adoption of the automobile by the average American consumer, America's roads were, to a modern observer, unfathomably bad. Outside of cities, they were almost exclusively dirt. For most of these roads, the only improvement over the wagon trails they replaced was that they were graded—that is, smoothed by a wooden board dragged by horse or ox. After rainfall they became muddy or even impassable. One of the reasons for the profound isolation of rural communities at the time was that it was barely possible to get anywhere that rail lines did not go except on foot or on horseback.

It wasn't until 1903, the same year the Wright brothers flew at Kitty Hawk, that anyone managed to complete a transcontinental automobile trip. (Others had tried and failed.) It took sixty-three days and was considered so incredible that crowds gathered whenever Horatio Nelson Jackson, Sewall K. Croker, and their bulldog, Bud, who wore his own pair of road goggles, rolled into town. On their way from San Francisco to New York City, the trio were punished by deeply rutted roads pockmarked with divots and the occasional tree stump. Their vehicle broke down frequently, forcing them to wait for replacement parts to arrive by train.

When they finally made it to New York City, the *New York Herald* of July 27, 1903, reported that "the mud besmirched and travel stained vehicle which had borne them so faithfully and sturdily over fifty-six hundred miles of roads between the Pacific and the Atlantic . . . was visited by admiring automobilists, and curious passersby peeped in upon it. In honor of its achievement it was decorated with tiny flags and draped with national standards."

At the time, the only paved roads in the country were clad in stone. Freight-hauling horse-drawn wagons, which were the heavy trucks of their day, necessitated their construction. It was far too expensive to pave the rest of the country's roads in the same manner, but a growing demand by bicyclists and then hobbyist owners of automobiles resulted in a rapid technological evolution that would eventually have as big an impact on America as the development of germ theory and steam engines.

From graded dirt roads, civil engineers upgraded intercity roads to a mix of sand and clay, which still broke down under heavy loads. Gravel roads were a little better, but they required constant maintenance and were particularly ill-suited to the narrow tires of early cars and trucks.

In the early 1800s, a Scotsman by the name of John Loudon McAdam hit on the roadbuilding technique that still bears his name—macadam. In his method, smaller stones were compressed by a heavy roller into a tight mass that held up better than anything aside from actual paving stone. Unfortunately, even his improved gravel roads required constant maintenance.

The invention of modern road surfaces was in some ways a happy accident. In America, roadbuilders of the first decades of the twentieth century spread asphalt, a petroleum by-product that bubbled out of the ground naturally in the newly exploited oil fields of Texas, on top of their rock and dirt roads in order to keep dust kicked up by passing wagons and vehicles from choking passengers. The result, a combination of tar and macadam, is what we now know as tarmac.

All of this evolution was required for builders to get us to a point at which a road surface could be tough and smooth enough for automobiles, and especially trucks, to roll over it at high speeds. Just as roads were evolving, the number of trucks in the country was exploding. Through 1910, fewer than 10,000 trucks had been produced in the United States. In 1911, U.S. factories churned out 13,319 trucks. By 1913, that number had quadrupled. World War I was fought on foot and with the aid of horse-drawn cannon, but by World War II, the U.S. Army consumed 2 million trucks.

It was more than enough to rescue the floundering Cummins diesel engine company from bankruptcy. Today, Cummins produces more diesel engines for trucks than any other company in the United States. It's the engine in Robert's truck, almost certainly the one in the truck hauling our box across the country, and the source of that distinctive rumble you've probably heard a thousand times.

City planners continued to plot ever-wider roads, but they all suffered from the same essential problem, which was that they were just expansions of the same roads humans had built since the first of our bipedal ancestors decided to widen a game path by treading on its shoulder. Roads into which driveways and pedestrian crosswalks could spit traffic were not places where automobiles, which by now could already achieve speeds in excess of sixty miles an hour, could be used to their fullest potential. As trucks and automobiles rapidly evolved, it became clear that, beyond what they were made of, the way roads were laid out needed to be rethought. The goal was driving without impediment, which would require keeping pedestrians, cyclists, children, horses, and automobile cross-traffic out of the way.

In 1930, the same man who was responsible for the Appalachian Trail, conservationist Benton MacKaye, published an essay on the "townless highway" in which he proposed roads that would be limited-access, to separate them from towns. The idea was that towns would be saved from the noise and bustle of fast-moving automobiles, but the eventual reality—elevated highways devastating the civic life of cities all across America as they plowed through historic neighborhoods—would lead him to regret his invention.

This is where Hitler comes into the picture. As with so many other innovations that were in the air but had yet to be fully realized, including lightning war, rockets, jets, and nuclear bombs, Hitler saw the highway as a way to make Germany so powerful that its rule would last a thousand years. Some economic historians argue "Hitler's Highway," what is now known as the Autobahn but was once known as the Reichsautobahn, was an important stepping-stone in his consolidation of power. One of his first mass building projects, it was a showpiece that served as both a testament to German engineering and a gigantic make-work project for tens of thousands of unemployed German men.

Begun in 1929, the Autobahn was adopted by Hitler almost immediately after he came to power. By his own account, he saw its potential as a tool for swaying reluctant working-class Germans in his favor. In the nine months between the November 1933 parliamentary election, the first since the Nazis seized power and banned all other political parties, and the August 1934 referendum granting Hitler supreme power, at least one in every ten Germans switched from opposing to supporting Hitler's regime in areas where the Autobahn was under construction. By 1941, when construction was halted because war needs took precedence, Germany had completed almost 2,400 miles of the Reichsautobahn. At the time, few Germans owned cars. The highway was entirely a showpiece and a weapon of war, ending as it did on Germany's borders with France, Belgium, Poland, and the like, allowing rapid transport of supplies, tanks, even pieces of U-boats to those borders.

The world's first superhighway system was a unique enabler of Germany's blitzkrieg. Ironically, near the end of the war, the Autobahn made a significant difference in the Allies' ability to chase German troops across the country, by facilitating the movement of supply trucks that had taken a beating on the rough roads of France.

"After crossing the Rhine and getting into the areas of Germany served by the Autobahn . . . our maintenance difficulties were over," wrote E. F. Koch, a highway and bridge engineer with the Ninth Army who would later work for the U.S. Public Roads Administration. "Nearly all through traffic used the Autobahn and no maintenance on that system was required."

The Autobahn, with its limited access ramps, overpasses, clover-leafs, divided lanes, and generous shoulders made it the world's first fully modern highway. Other highways of comparable engineering and similar vintage, like the Pennsylvania Turnpike, did not match its scale and ambition. In the conception and construction of almost all early highways, the countless bureaucrats, politicians, pundits, engineers, and business groups involved were concerned chiefly with the movement of automobiles. They did not appear to realize that what they were really building was the greatest network of freight transportation ever conceived.

The Autobahn made a deep impression on Eisenhower, who immediately after the war was the military head of the U.S.-occupied portion of Germany. He later cited an experimental 1919 truck transportation convoy he had been a part of in his early years in the army—a rough journey on mostly unpaved roads from Washington, D.C., to Oakland, California, that took fifty-six days—and the sharp contrast of the Autobahn as motivation for finally getting a U.S. highway system funded.

"After seeing the autobahns of modern Germany and knowing the asset those highways were to the Germans, I decided, as President, to put an emphasis on this kind of road building," Eisenhower later recalled. "This was one of the things I felt deeply about, and I made a personal and absolute decision to see that the nation would benefit by it. The old [1919] convoy had started me thinking about good, two-lane highways, but Germany had made me see the wisdom of broader ribbons across the land."

And so was born our modern system of superhighways, officially known as the Dwight D. Eisenhower System of Interstate and Defense Highways. America now has 46,000 miles of interstates and a total of 164,000 miles of highway, composed of more than 1.5 billion metric tons of materials such as rock, sand, and cement. In addition to being almost inconceivably large, it's also the greatest public works program ever and one of the largest public subsidies of private enterprise in the federal budget. Taken as a whole, it is absolutely essential to the more than 3 *trillion* miles Americans travel by motor vehicle and the 12 billion tons of freight transported by truck every year.

All of that money, time, and effort were expended in order to accomplish this straightforward goal: reduce the rolling resistance of the wheels of a truck to the point that goods can be carried from any city, town, or hamlet in America to any other. It's this system that I find myself suddenly unable to take for granted as Robert finally drops me off in Derby, New York. Afterward, he heads to the nearest truck stop to find a place to sleep. Meanwhile, I'll find myself the sole guest in a forty-six-room mansion that was recently converted into an inn that appears to suffer, as do many businesses in small towns bypassed by interstates, from a permanent lack of traffic.

▶ ▶ ▶

Day two of the journey of our USB charger across America sees it cross the Rocky Mountains and the Continental Divide, that crest in the United States from which all water flows either east to the Atlantic Ocean or West to the Pacific Ocean. On I-70, this means falling into line with large amounts of other semitruck traffic as drivers strain to make it up to the Eisenhower Tunnel, which passes under the Continental Divide and at 11,000 feet is the highest tunnel in the entire interstate highway system, as well as one of the highest in the world.

Other than ice and reckless drivers, there is no bigger threat to a semitruck than rolling down a long, steep grade, and slopes of up to 6 percent can be found on both sides of the Eisenhower Tunnel. Inexperienced truckers feeling pressure to maintain their speed who don't get into a low enough gear at the start of their descent can burn out their brakes, leaving them no ability to stop at all, unless they're lucky enough to maneuver onto one of the highway's runaway truck ramps without flipping their truck first.

Since the 1950s, trucks have had "Jake brakes," which allow them to use their engines to slow themselves on such descents, but these too have their limits. Mountain passes require skill and try even the most experienced truckers.

East of Denver on I-76, hundreds of miles of mostly downhill help our charger make up time as it cruises into the vast agrarian flatness of Nebraska. Stopping for the night just outside the state capital of Lincoln, one tiny gadget and 60,000 pounds of other goods are finally within striking distance of the Amazon distribution center that is their ultimate destination.

▶ ▶ ▶

The morning after my ride-along with Robert, I find out just how punishing a business trucking can be even for those who own the trucks that drivers find themselves bound to.

Ray Kaczar, Robert's boss, has spent forty years as a truck driver, off and on. He owns a six-truck company called Avalon National, based in Cassadaga, New York. He has white hair, glasses, and a brand-new Dodge Ram truck. He wears button-down shirts. We talk over lunch at a locally owned truck stop, one he prefers because it treats drivers right. "Not too many left like this one," he says as we sit down in a booth. Outside, a letterboard sign says BEST LIVER AND ONIONS.

Ray remembers what it was like before deregulation, when a solo trucker could make enough to buy a house and a boat and still save for retirement. He weathered the oil crisis and capitalized on the industry's post-deregulation transition to cutthroat competition by starting his own trucking company. At one point he owned a company with fifty trucks. After selling the company, he had a go at being a farmer, which turned out to be even harder than trucking. At sixty years of age, he hit the road as a solo trucker once more.

In 2018, Avalon was named a Carrier of the Year by Eden Prairie, Minnesota–based freight broker C.H. Robinson (CHR), which is the biggest company of its kind in the world by volume. Freight brokers are the middlemen between shippers and trucking companies. They are the aggregators of demand from shippers and the matchers of that demand with the capacity of trucking companies, down to the level of individual drivers and their trucks. Ray is fiercely loyal to CHR; he says they take care of him and his drivers and don't screw him around like other freight brokers have.

In theory, a freight broker is merely an owner of a market in which shippers and truckers with empty trailers are matched, the sort of thing a company like Uber accomplishes in an analogous marketplace that matches drivers and passengers, using nothing but computers and algorithms. In practice, freight brokerage is a deeply relationship-based business in which shippers are not only evaluating the cost of moving a ton of freight but the odds that it will arrive on time and intact. For manufacturers, materials or parts that are late can be as bad as ones that don't arrive at all. For retailers, getting goods to a store in time to avoid dreaded out-of-stock situations is critical to both maximizing sales and keeping customers coming back when they have the choice of shopping elsewhere, or online.

Despite its size, CHR, like nearly all freight brokers, is still very much about people. A single agent handles all of Avalon's truckers, getting to know each one individually. She pulls individual runs from CHR's internal freight board, where shippers post what kind of freight they need moved where and when. She's chatty, personable, and knows the names of the dogs and spouses that ride along with Avalon's drivers, her charges. They trust her, and she is so integral to the way Avalon functions that Ray often refers drivers to her when they have questions about the details of a load they'll be carrying.

Ray is deeply spiritual, an idealist, can't stand the sight of a home-less person, and thinks it's shameful that Americans, children of God all, have shirked their duty to take care of one another. All his drivers, he says, have been "used and abused" by companies who lied to them or even, in some cases, stole their pay. He collects them, puts them in trucks, does what he can to earn their trust. Robert says Ray is a good boss but that he might eventually have to quit anyway, since Ray keeps putting him on loads that take him too far from his home in Orlando. Ray says it's hard to find loads into and out of Florida and that both Avalon and Robert are at the whims of the freight market.

"One thing we have that I have never seen in my life [in a truck-ing company] is we are drama-free," says Ray. He's parted ways with drivers who didn't fit into the kind of workplace he's trying to create. "They were good drivers, but you don't need that negative energy," he adds.

Ray's dedicated account manager at CHR makes sure all of his drivers are picking up loads almost as soon as they've dropped one off, a tricky process essential to the profitability of all trucking companies and the freight brokers that sell their capacity to shippers. Even so, Ray lost $5,400 in the month before we spoke, as shipping prices slumped across the industry, a product of a softening economy in late 2019 and growth in the number of small trucking companies like his.

In an economy in which bigger always seems to be better, mergers and acquisitions are the norm, the total number of listed stocks continues to shrink, and economists knit their brows over the disproportionate gains of superstar firms capturing markets and transforming their dominance into an unassailable lead in technology and capital, trucking remains a remarkably fragmented industry. The biggest private fleets are in delivery to homes and businesses—think PepsiCo, not Walmart—and the largest independent trucking companies have no more than a single-digit percentage of the overall market. CHR, one of the few behemoths in the space, doesn't even own trucks, but sells their services to shippers. The majority of the trucking capacity it offers is from small- and medium-size companies with less than a hundred trucks.

I ask everyone I encounter in the industry why long-haul trucking is so resistant to consolidation. It appears to boil down to this: trucking is an industry where, at the level of companies that own trucks, there are no advantages to scale. Instead, there are liabilities that can mean the end of a firm as soon as the economy turns sour and demand for trucking falls. Those liabilities include capital tied up in vehicles that can cost upwards of $300,000 apiece, as well as enormous fixed costs, including thousands of dollars a month in licenses, insurance, fuel, and maintenance, not to mention wages for truck drivers, who despite their historically low compensation are still the single biggest expense in moving freight via truck.

The average small- or medium-size trucking company is lucky to break even, much less make a profit, and Ray's is no different. He says he has a handful of drivers eager to start driving for him, but "I need another truck like I need a hole in the head." Only in trucking would an award-winning and beloved small business with practically unlimited demand for its services hesitate to expand.

Some of the struggles of truck companies are due to the fact that shippers have a tremendous data advantage over truckers. A number of platforms allow shippers to see what rates their competitors are paying. Even when they can't, online marketplaces for freight shipped by truck give them a sense of the going rate. This leads to a form of "monopsony" in which the buyers of truckers' services have all the power, and truckers, the overwhelming majority of which no longer belong to a union, have little or none.

In a monopoly, there's just one supplier of a good or service. In a monopsony, there's just one buyer. The classic example is a company town belonging to the owner of a mill or mine, where there is only one employer. That employer can set—and depress—wages accordingly. In the case of the trucking industry, it's as if shippers, acting as one, are the sole employer of trucking companies, which are numerous and fragmented, leaving them powerless to demand higher rates.

While this situation keeps the cost of shipping goods on trucks low, it has led to massive inefficiencies in the industry as a whole.

The most obvious is the amount of time that trucks, each one a hugely expensive piece of machinery, sit idle. For comparison, if ocean-freight companies ran their business the way the trucking industry does, ships would sit idle at ports for half their working life. That's a situation the highly consolidated and technologically sophisticated container shipping industry has done everything it can to avoid.

Much of this idle time is due to shippers being slow about loading and unloading trucks, one of the few problems everyone in the trucking industry agrees needs to be fixed. Truckers and trucking companies have no power over this part of their day, other than to refuse to deal with shippers who abuse them. For a variety of reasons, this is not a very effective threat. Among them: truckers will take loads they lose money on if they get them home or to their next, more profitable, pickup; small trucking companies don't want to strain their relationships with the freight brokers that sell their services to shippers; and inexperienced truckers who are paid by the mile may be in what is essentially debt servitude to big trucking companies as they pay off the cost of their own driver training, leaving them little choice about anything.

"We have enormous inefficiencies in this system," says Steve Viscelli, a sociologist who studies the trucking industry. "It's just that workers essentially pay the price of them because they own the asset that's underutilized or they're the ones not getting paid."

In an average day, a driver sits unpaid, doing nothing, for hours while their truck is loaded. Then they drive it seven to eight hours and wait again while it's unloaded. Finally, they drive it to a truck stop to use it as a motel for ten hours.

"Can you think of another expensive piece of equipment used in this way anywhere else in the economy?" says Steve.

In 2018 and again in 2019, venture capitalists poured about $3.6 billion into trucking and related freight start-ups, an amount without precedent in the history of the industry. Marquee names like Uber are using investor capital to build Uber Freight, which claims to eliminate some of the inefficiencies of trucking by, for example, having drivers drop off and immediately pick up fully loaded trailers rather than waiting for them to be emptied or filled.

But all these efforts should be regarded with a healthy skepticism. Truck transportation is the land war in Asia of investment opportunities. Ray's own experience with start-up Coyote, a logistics firm recently acquired by UPS, and Amazon Logistics indicate that smartphone-accessible, Uber-style marketplaces for booking truckloads of freight may not gain much traction in an industry in which every driver and trucking company is an eccentric combination of needs, wants, short-comings, and talents, and best suited to certain regions, conditions, times of day, types of freight, and shippers.

For Ray, the new style of online-only freight marketplaces that purport to cut out the middleman of human freight brokers and agents is fundamentally incompatible with treating his drivers fairly. A canceled ride on an Uber means a minor inconvenience for a driver who may be one of thousands in a city, accepting dozens of rides a day.

A similar snafu for a trucker means the misallocation of an asset that costs its owner hundreds of dollars an hour even when it sits idle.

Ray quit Amazon Logistics after just a couple of weeks. "They're liars," he says. "They gave me one load, a round trip pulling their trailer, and it paid good money. And as soon as they assigned that load to me, another one going on the same run but with a different trip number attached to it popped up, and for that one they posted a much lower rate. Someone must have taken that lower rate, and then they didn't even call me. They just emailed me: 'Your load's been canceled.' That's the ethics of Amazon."

▶ ▶ ▶

From Nebraska, it's an early-morning dash for our box through Iowa, on I-80. In Des Moines, Iowa, after a hard left turn northward onto I-35, the truck is running flat out, the driver going as fast as he can without incurring a ticket, since he has to arrive within an hour of his appointed time at Amazon's fulfillment center in Shakopee, Minnesota, a suburb of Minneapolis.

Delivery at Amazon fulfillment centers is professional but alienating, says Desiree Wood, a trucker and the founder of Real Women in Trucking, which advocates for the approximately 6 percent of truckers who are women.

At the gate, drivers are handed something resembling the buzzer you get at a restaurant to let you know when it's your turn to be seated, but other than that, no one talks to them. As they back the truck into the loading dock, they can't even see workers inside the distribution center. Even the receiver at the dock is, for their safety, in a sort of cage.

Compared with the best shippers, the entire process of dropping a trailer at an Amazon fulfillment center is so impersonal as to be "kind of degrading sometimes, how you're treated," says Desiree.

"No one says, 'Hey, thanks for being here, there's a lot of people who are really going to be excited that their order has arrived.'"

Amazon is definitely not the worst, she adds. Many truckers I spoke to cited food distributors as the most likely to be rude to drivers, over-charge them for loading and unloading their trucks, or waste their time.

Trucking, in its hunger for more and more drivers, and despite its enthusiastic embrace of technology, demonstrates something fundamental about logistics in general: it remains an irreducibly labor-intensive industry. In a world in which adding IT, AI, and robots to any enterprise is supposed to squeeze the humanity out of it, trucking shows how the need for what humans do best—adapting to unpredictable conditions, problem-solving in the face of mechanical and technological failure, getting creative in situations in which the rules really aren't set up for anyone to succeed—is more pressing than ever.

There are forces that could fundamentally transform truck trans-portation. In the long term, they are what many of those $3.6 bil-lion worth of freight start-ups are betting on. Fully autonomous or semiautonomous trucks have the potential to upend the economics of trucking by eliminating or supplementing the most expensive part of the industry. They could also do things no human can accomplish, such as running a truck 24/7.

Chapter 13

THE FUTURE OF
TRUCKING

I'm cruising down a desert highway at seventy miles an hour in a twenty-ton semitruck with more processing power than the world's fastest supercomputer possessed as recently as a decade ago.

Four of us share the cab of this truck, but no one is driving. We are all just passengers. In a worst-case scenario, our vehicle has enough kinetic energy to turn a passenger car stopped on the road ahead into an airburst of steel, plastic, and human confetti.

Controlling this truck is a computer that's dreaming, in a way—a dream not so different from the one you're experiencing at this very moment, and in every moment of your waking life.

Our waking dream, and the truck's, is a fable about the state of our bodies, the world outside, and possible futures both might soon inhabit. The truck is using deep learning to construct a reference view of reality, including the state of the road ahead, the vehicles around us, and the tarmac beneath our eighteen rumbling wheels, each spinning sixteen times a second. Within 100 milliseconds, the truck's AI needs to digest all its sensory inputs by converting pixels captured by its cameras, and a point cloud generated by its lidar, into cars, trucks, and motorcycles. Then analyze the pattern of their motion and predict their future behavior. Then localize the truck itself in space by matching what it sees against what is stored in its high-resolution map. Then determine its next action, and very precisely control its throttle, brakes, and steering.

The AI driving this truck is a nearly production-ready, fully autonomous driving system designed and built by TuSimple, a San Diego–based firm with offices in Beijing that is aggressively pursuing the American trucking market. The truck's "brain" is using various means of prediction, principal among them machine learning technologies including neural nets, to guess about the future behavior of the humans around us. Humans, the ultimate uncontrolled variable, are by far the biggest challenge and the greatest peril faced by our artificially intelligent driver—and its passengers.

This autonomous self-driving system relies on a mental model of the world around it far more complete than the ones employed by even the most safety-obsessed human drivers. TuSimple's engineers will tell you they built their systems this way out of an abundance of caution, but it's also because of how dependent on context modern AI can be. It's what AI expert Gary Marcus calls brittle and shallow— that is, not terribly adaptable, and lacking what humans call common sense.

All self-driving AIs require huge amounts of data specific to the type of vehicle they will pilot, and the context in which it will take place.

All these are reasons to give the system both superhuman senses and intricate knowledge of the roads on which it will drive. The latter takes the form of a map unlike any created before it. Unbelievably detailed, three-dimensional, overlaid with dozens of layers of metadata, it recalls Jorge Luis Borges's mythical map that was as big as the empire it was intended to represent. Centimeter-precise, this map is so good that a human watching it unfurl on a monitor could use it to pilot this truck even if all of its windows were painted black.

Making these kinds of maps is an industry unto itself. Companies like Waymo, a subsidiary of Google parent company Alphabet, create their own using a combination of sensors to assemble and frequently update them. These maps include the boundaries of lanes, the height of curbs, the speed limit both posted and generally accepted, and every last road sign, traffic light, crosswalk, and potential hazard.

The AI on board TuSimple's self-driving truck draws about four times as much power as the average American home and performs hundreds of trillions of mathematical operations per second. About a third of its computational wattage is devoted to map-related tasks.

Step one in translating our truck's perception into action is determining precisely where we are on this map and then updating this information dozens, even hundreds of times per second.

This turns out to be a fantastically complicated problem.

"We use IMUs, GPS, SLAM, visual, lidar, radar," says Chuck Price, the white-haired and well-seasoned chief product officer of TuSimple, who is riding with me in the sleeper cab of the truck. In front of us is the truck's human "driver," an experienced trucker whose hands aren't on the wheel. Riding shotgun is the safety engineer, who has a laptop perched on his knees. We are traveling southeast on I-10 near the company's primary U.S. testing facility in Tucson, Arizona. "All that comes together to give us a very precise view of where we are in the world," Chuck adds.

The phrase "comes together" does a lot of work when describing the integration of this many sensors, each with its own capabilities, strengths, weaknesses, and peculiar histories.

As the flat, remarkably straight expanse of I-10 flows by in front of us, I recall that IMUs, which stands for inertial measurement units, have something to do with guided missiles but are also present in our smartphones.

Later, my research reveals that the IMU in this truck is yet another example of technology that Nazi-era Germans either invented or greatly advanced, and which has subsequently been pressed into service for our everyday convenience. (Other examples of such technologies include, of course, the highway on which we are riding and also, contrary to a view widespread in the Anglophone world, the world's first digital computer.)

The earliest iteration of the IMU took shape when Robert Goddard, pioneer of modern rocketry and inventor of the world's first liquid-propellant rocket engine, used gyroscopes to aid in the steering of one of his rockets, launched in New Mexico in 1932. His ideas were elaborated by Wernher von Braun, whose V-2 rocket production program in Nazi Germany worked to death more than 10,000 slave laborers but did not, in the eyes of the U.S. government, outweigh his potential utility to America in winning various Cold War races against the Soviets. (At the close of World War II, Braun was shipped to the United States as part of Operation Paperclip and would prove instrumental in helping capitalists beat communists to the moon.)

On the V-2 rocket, Braun gave the world its first real inertial navigation system, as it was then known. One of the "vengeance weapons" that Hitler thought would be so devastating and terrifying that it would force the Allies to call off their attack and press for peace, the V-2 was primitive by today's standards but effective. Its biggest issue was that it wasn't terribly accurate. Its inertial navigation system could do little more than keep it on a particular compass heading—in this case, toward London.

In parallel, Charles "Doc" Draper, an engineer who would go on to become legendary at NASA, worked during World War II on a gyroscope-assisted gunsight to allow U.S. Navy gunners to fire shells accurately despite the rocking and pitching of a ship. He was inspired by the instruments on early airplanes, which also relied on gyroscopes. Appropriately, his invention later evolved into an inertial navigation system for airplanes and made possible in 1949 the first ever autonomous flight of an airplane. Later, a much-refined version of this system flew on the Apollo spacecraft and was essential to getting humans to the moon.

An IMU of the sort invented by Draper measures rotation, using gyroscopes, and motion, using accelerometers. It integrates the resulting data using a computer and can determine the position of an object in space by dead reckoning. Dead reckoning is the process of determining one's location by first fixing one's position relative to known points and then tracking your speed and direction of travel— an important navigation technique for sailors for at least the past 500 years.

IMUs are now used in rockets, satellites, airplanes, drones, smartphones, smartwatches, and, of course, the autonomous driving systems in vehicles of every kind, including TuSimple's trucks. Doc Draper's first ever IMU, tested in 1949, weighed 3,000 pounds. One put into guided missiles in the early 1960s weighed about 120 pounds. A much-refined version of that IMU that flew on the Apollo 11 spacecraft in 1969 was the size of a volleyball and could be lifted by one person.

Fast-forward to the present, through the commercialization of the integrated circuit and countless subsequent generations of Moore's law, and the IMU in your phone or smartwatch has shrunk until it's now about the size of a sesame seed—three millimeters square and 0.6 millimeters thick. It's etched out of silicon and has parts that are identifiable as an integrated circuit. But its operative microscopic bits are something other than an integrated circuit—they're actual, movable pieces, known as microelectromechanical systems, or MEMS. MEMS are one of the minor miracles of modern photolithography, the technique used for etching microchips from silicon, which is probably the single most important engineering breakthrough since mass-produced steel.

To sum up, every time you rotate your phone 90 degrees to watch a video or take a picture and the phone successfully recognizes its change in orientation and pivots its interface appropriately, you're relying on a device the size of a large grain of sand, perfected by engineers who wanted to make sure our nuclear-tipped missiles found their mark inside Soviet Russia. (Also, some of them used to be employees of Adolf Hitler.) At some point in the not-too-distant future, as a part of the positional system of a vehicle, the accelerometer that recognizes the orientation of your phone is going to be an integral part of a system that makes it possible for you to take a nap while your car cruises at highway speeds.

GPS, or the Global Positioning System, is a natural complement to an IMU, since all methods of navigation by dead reckoning must begin with a known position and GPS can provide that initial fix.

As a form of celestial navigation, GPS has a heritage far more ancient than inertial navigation, only in place of fixing one's place relative to stars in the night sky, it uses human-made celestial objects. GPS depends on an array of satellites that orbit Earth twice a day, traveling through the near vacuum of space at 8,724 miles an hour at an altitude of 12,550 miles.

Every time your phone gets a GPS fix to help you navigate to your next destination, it's a combination of wayfinding—pioneered by Polynesian maritime navigators, who used it to travel thousands of miles with no technology more sophisticated than their powers of observation and a detailed mental map of the positions of stars and the sun—and Einstein's discovery of relativity. On account of traveling so quickly and the reduced gravity at their altitude, the precise atomic clocks on board each of the world's twenty-four GPS satellites tick just a bit slower than their counterparts on the ground, which is a relativistic effect predicted by Einstein. If we didn't know about it, and those satellites weren't equipped with electronics to account for relativity, the entire system would cease being useful in about two minutes.

SLAM stands for simultaneous localization and mapping, and it's a way to integrate all the various sources of information about the outside world and a vehicle's place in it. In TuSimple's trucks, the mapping part of SLAM is carried out by the high-resolution map. The resulting system is easiest to understand in terms of ourselves, because as humans we do it all the time. Whenever we enter a familiar environment and immediately begin building a map of it from the information entering our eyeballs—both our "flat" view of the scene and the information we gather about the distance to surfaces and objects—we're doing something analogous to what a self-driving vehicle must do.

This system works much better when fed additional sources of information about the shape of objects and their distance from us. Those other sources of information can include depth sensors that beam out a grid of dots invisible to us but visible to some cameras. For most phones' front-facing facial-recognition sensors, this information is acquired by shining an array of infrared dots at your face. The distortion in that grid tells the sensor in an infrared camera about the shape of whatever object it's looking at.

For most self-driving vehicles, something similar is accomplished with lasers, and it's called lidar. In lidar, rather than looking at the distortion in a grid of points of light, rapid pulses of laser light are fired at objects in front of the sensor, and the amount of time it takes the light to return tells the sensor precisely how far away that object is.

The truck's AI takes information from all of these sensors and creates a map of the environment and our truck's position within it. A similar process is integral to any attempt to overlay a fictional reality atop our real one, as in applications of augmented reality (AR). If you've ever played with an AR app on your phone—say, the Ikea app, which shows you what furniture would look like in your actual home, using your phone's camera and screen—that's accomplished largely through SLAM.

TuSimple's system relies on "sensor fusion," as the last step of localizing the truck in space. In the continuum from data to knowledge, sensor fusion is the part of perception in which information from all of a machine's—or our own—senses is collected and synchronized. It's essentially the creation of an internal consensus reality. For the purposes of a self-driving truck, sensor fusion is the process by which data from all those different sensors—the IMUs, GPS, cameras, lidar, and radar—are brought together in a single virtual world that depicts the moment-to-moment reality around the truck.

And it's within this dream of reality, the same one you inhabit right now, that the truck's artificial intelligence actually learns and acts. This simplified, averaged, and partly inferred simulation of the outside world—a blend of what's sensed in this instant with things remembered from a past encoded in the system's detailed, 3D map of the world—is an essential buffer between the part of the machine that makes decisions and the parts of the machine that sense what's going on in and around the truck. On the surface, at least, the virtual world presented to the truck's AI has remarkable parallels with what goes on in your head.

One reason this "digital twin" of the world outside is so important is that it allows the truck's AI to interact with a relatively stable and highly accurate version of reality, with almost all the errors, which are inevitable in individual sensors, filtered out. The human brain does something similar, and what happens when this function of our minds breaks down is telling. Schizophrenia, for example, derives in no small part from disorders in sensory processing. Creating an accurate, or at least useful, view of the world around us is one of the most essential functions of a brain, whether biological or digital.

On TuSimple's truck, this interpreted reality is displayed on monitors visible to those riding along. They depict a third-person, three-dimensional view of the truck we're in and the road ahead of us and to each side. It looks like a video game, with the white avatar of our truck, the protagonist, always in the center of the screen. The boundaries of the road are depicted in cool blue lines, and everything that isn't salient to safe operation of the vehicle fades into a black background. The overall aesthetic evokes the world inside the computer in the Disney movie *Tron*.

This view of what the truck perceives is essential not only to its own AI but to the humans who ride within the vehicle. It's a map within a map, showing what the truck sees and intends to do at any moment. This transparency, this view inside the mind of the machine, is essential to how the truck is being monitored and trained by a team consisting of an experienced truck driver and a tech-savvy safety engineer.

Today, our driver is Brett Bernard, a man in his fifties who has been driving for thirty-two years. Before working for TuSimple, he was driving a chemical tanker, carrying mostly acids from Arizona to Houston, or even as far as New Orleans. Carrying hazardous liquids is one of the most challenging, dangerous, and highest-paid jobs a trucker can have, and it requires a great deal of experience and a nearly spotless driving record. Brett found TuSimple through word of mouth—a buddy was already driving for the company—and in less than a year on the job he's already logged 20,000 miles.

In the passenger seat, with a notebook computer in his lap, is John

Panttila, the safety engineer. John, who goes by JP, is in his forties. He has a background in civil engineering and wastewater, but about twenty years ago he got a commercial driver's license. He says he never expected to combine his engineering and truck-driving backgrounds. He was the company's first safety engineer in Tucson, and he has trained all of those hired since.

JP's enthusiasm for the job is considerable, even a bit cultish, something I've seen before in people who have devoted their lives to a start-up and have toiled and sacrificed to make it grow.

"I believe in what we're doing—the cause of what we're doing is what's drawn me to be here," says JP. That "cause" isn't replacing humans in trucks altogether, he adds, which surprises me. Rather, he continues, it's about making trucks safer by supplementing the faculties of a human driver.

I ask Brett about the explicit goal of TuSimple as a start-up, which is as much about getting truckers out of the cab altogether as it is about making human-driven trucks safer. Does he feel threatened? Do any of his trucker buddies razz him about helping to train the technology that could someday put them out of a job? He deflects the question, says it never comes up, that he's just happy to be here. TuSimple treats him well, he adds.

Throughout our drive, JP intermittently directs rat-a-tat verbal outbursts at Brett, like a cross between a livestock auctioneer and a Navajo code talker, speaking a language known only to the two of them. At one point JP says, "Lane 3 is back out in ninety meters—" and then he's cut off by the loud, female, unmistakably artificial voice of the truck's AI: "CANCELED," she declares.

Throughout the drive, JP the engineer is head-down in front of his laptop, telling Brett the driver what the AI is seeing on the road ahead. Brett is the one person in the truck who is not looking at a screen. The purpose of JP's cryptic rambling is to let Brett know what's going on in the truck's AI—not just what it perceives but, just as important, what it intends to do about it. Will the truck soon speed up, slow down, change lanes? Only JP, eyes glued to his laptop, knows.

Downstream from the truck's map- and sensor-based view of

reality, but also in constant feedback with it, is the part of its system that decides what to do with all that information. In this part of the truck's "brain," there is a great deal of machine learning and AI as well. Most of the truck's behavior is determined by this AI, based on a large volume of statistical data about best driving practices, which is collected by TuSimple's trucks, while humans pilot them, and is then curated by the company's engineers. Those engineers tune this data and the algorithms that use it, adjusting priorities and other parameters to refine the truck's behavior.

Outside, the sun is painfully bright and the sky overwhelmingly blue. Inside the truck, it's so cold that Brett is wearing a jacket, and I'm beginning to regret not dressing for the eternal fall of Tucson's indoor climes.

Exiting the highway, we come to a particularly tricky intersection, a highway overpass where we'll turn around. As the truck slows, we're treated to a prime example of how the truck must explicitly obey the instructions of its human masters, rather than learning how to handle this intersection on its own, which is the sort of thing you can allow an AI tasked with identifying cats to do, but not one that's in charge of a twenty-ton truck.

Where the exit ramp we're on ends, the street is unusually narrow and bends sharply to the left. The result is that any vehicle sitting at the light at the top of the exit ramp can't see around the concrete barriers there in order to determine whether traffic is oncoming.

The logic of dealing with this strange intersection, an artifact of when this stretch of highway was far more remote and traffic engineers didn't imagine any trucks would use it regularly, is not the sort of thing engineers can just leave an AI to absorb by watching humans handle it. On the monitor showing what's going on in the truck's "mind," a colorful pattern of overlapping stripes in red and yellow shows what the truck should do and what it should avoid. Getting through this light and the one just past it, which has funny timing and gates the entrance of vehicles onto the overpass proper, requires that the truck keep rolling even when it might normally stop.

This situation is a good illustration of the frequently imperfect and entirely handcrafted world that autonomous vehicles must navigate. We all know that one intersection in our neighborhood where a bush obscures our view of traffic crossing our path, or the one where there should be a light, or that other one where people tend to take the curve too quickly and accidents happen all the time. To make it safe for autonomous vehicles to operate on our strange and inconsistent roads, we have to teach them not just how to drive under optimal conditions but how to handle the peculiarities of our built environment, just as we might teach a teenager just learning to drive.

And if the AI can't handle it, well, that's what the human in the driver's seat, his hands off the wheel but always hovering near it, is for.

"We rarely have to take over, but we know that there are places where, because it's so complex, there's a chance we'll have to," says Chuck.

As we roll through the intersection, yellow light successfully ignored, many other algorithms are dumping data into the truck's view of reality, adding labels and context and imbuing with significance the things it perceives.

In the truck's perceptual system are the classic object-recognition algorithms, made possible by deep learning, that allow it to identify people, obstacles, and other vehicles. It's the same sort of technology that allows Facebook to recognize your friends' faces in photos, or Google to cough up pictures of felines when you type in "cats." Then there are the many niche algorithms the truck is running, each purpose-built. One discerns what traffic lights at intersections are conveying; another interprets turn signals on cars.

To interpret the truck's raw sensory input, its algorithms rely mostly on deep learning. One level above that, in the middle of the truck's complicated algorithm of interpretation and decision-making, it uses Bayesian analysis. Bayes' theorem is remarkably simple to write out, teachable in an hour by a skilled educator, and has endless implications for how decisions can and should be made. At its simplest and most abstract, Bayesian probability is a way to take what just happened and update your understanding of the state of the world. It's a way to make judgments about the likely status of a particular variable, or a whole world, based on very little information, and then continually update one's view of things as new information comes in.

In many forms of AI, one effect of Bayesian reasoning is plenty of information about how confident a system is in its own perceptions and judgments. This is useful to have when something new comes over the horizon, like yet another car or debris in the road.

The result is an accurate world model, a sensor-fused consensus of all those sources of data, a working hypothesis about the state of all that the AI can see, updated by the millisecond. For the truck, this means that "we're giving [the AI] a much richer description of the world than what's coming in from individual sensors, so that makes it far more robust," says Chuck. "It's not brittle. That's why we can plug a new sensor in and test it, and we don't have to rewrite a bunch of code."

TuSimple's target date for release of a fully autonomous self-driving system that does not require a human safety driver as a backup is 2024. The challenge, says TuSimple cofounder and chief technology officer Xiaodi Hou, is that for his trucks to become a product, his team must take the company's systems from something like 99.99 percent reliable to 99.9999 percent reliable.

On top of this is the challenge that all self-driving systems face: every additional decimal point of reliability costs as much in time, energy, and money as all the previous ones combined. It's as if, says Xiaodi, "we are very close to the speed of light. The closer you get to the speed of light, the harder it is to accelerate, so you need a huge rocket."

That "huge rocket" is all the collected and refined data required to assure the truck can handle every situation it could ever be reasonably expected to encounter, plus all the work put in by engineers to ensure that the whole system is reliable.

The result of all that effort is software that runs on the rack of servers in a minifridge-size gray cabinet I'm sitting next to, in the back of the truck's cab. Later, when I open it, a powerful odor wafts out: ozone and paint, the off-gassing of printed circuit boards, fresh from the manufacturer, warmed by their exertions. In total, those servers consume up to five kilowatts of electricity, and they include powerful servers with chips made by Intel as well as tuned-for-AI chips made by Nvidia.

Everything about this system is only possible because of a marriage, in the past decade or so, of new algorithms for deep learning and new hardware from the likes of Nvidia. It used to be the case that microchips became faster every year because the individual elements on them shrank, allowing companies to cram more of them onto a given integrated circuit while also increasing its performance. This process, known as Moore's law, is the reason a modern smartphone is far more powerful than the combined calculating power of every computer on Earth until the late 1960s.

Moore's law hasn't been doing so well lately, as engineers reach the physics-imposed limits of shrinking elements on chips any farther. (The tiniest details of the most advanced microchips are already smaller than a hundred atoms in width. Even when these features can be etched in silicon, electrons tend to do funny, quantum-weird things at that scale.)

Nvidia and many other companies, including Google, Baidu, Samsung, Apple, and the like, have figured out how to continue to increase the performance of their chips even as Moore's law has slowed. The trick is designing their chips to perform specific tasks, rather than continuing to design general-purpose chips capable of anything. As a result, the rate at which computers can perform the calculations critical for many AI applications is progressing faster than the rate at which Moore's law allowed general-purpose computers to increase in performance every year, for all the decades it reigned supreme.

The upshot of all this is that the AI part of the truck's system— the portion that is parsing data fed directly to it by the sensors and making other sorts of low-level decisions—is no longer the problem. The real engineering challenge, says Xiaodi, is all the millions of lines of code that must be written, by humans, to form the "driving model" of the truck, and enable many higher-level forms of machine learning and decision making.

In an autonomous system, he continues, many like to talk about how many "teraflops"—that is, trillions of operations per second—their system can compute. It's this measure by which the cabinet of computers in the back of a TuSimple truck is faster than the world's fastest supercomputer of a decade ago. But those teraflops, he adds, are really only available for certain tasks. And a lot of what the truck must compute can't take advantage of all of this processing power.

All is not lost, however. AI is broader than just the parts of it that have come into vogue in the past decade or so—that is, the learning algorithms built on "neural nets" inspired by their biological equivalents. It turns out there are all sorts of machine learning—older and deeper and more arcane areas of mathematical inquiry—that are no less useful on account of their vintage.

In a long, hard slog that leverages the full breadth of computer science, Xiaodi says his engineers are figuring out ways to take things that no one's ever done on Nvidia's silicon and get them to run on Nvidia chips a hundred times faster than they would on the part of the computer where they typically run, those general-purpose chips made by Intel or AMD. It's all about taking algorithms that would normally require that one calculation be complete before the next can be attended to and transforming them into the kinds of algorithms where a bunch of calculations do not, at least for a moment, depend on the outcome of one another.

This process is called parallelization, and if you picture a kitchen tasked with baking a hundred loaves of bread, it's not hard to imagine. In the old way, the kitchen had one oven, and everything was about baking bread faster by speeding up the operation of that oven—a quest with obvious limits. In the new way, you give the bakers in the kitchen a hundred ovens. But, if your task is in fact baking one really gigantic loaf instead of a hundred little ones, this process is much trickier to parallelize. Maybe you can speed up some other aspect of it—the kneading of the dough by a hundred sets of hands, for example.

If Xiaodi is right and TuSimple can arrive at full autonomy before Waymo, Tesla, GM, Volvo, and everyone else who is attempting it, or at least get there around the same time, the implications for his company are profound. Whichever of these companies succeeds, the implications for trucking are even more profound—although not, perhaps, in the ways that are immediately obvious.

TuSimple's goal is not to build a vehicle that can go anywhere under any circumstances. That's an end so unachievable with present technology that only Elon Musk regularly claims to be anywhere near reaching it. (And, to be blunt, almost no one who is deeply involved in this industry, outside of Tesla, believes him.)

TuSimple's goal is to build a vehicle that spends 90 percent of its time on the highway, an environment well marked and relatively predictable. It might handle a few extraordinarily well-mapped turns and surface roads as it exits the highway and pulls into the parking lot of a warehouse, the sort that Walmart uses to consolidate and redistribute goods headed for its stores, or that FedEx, UPS, and the U.S. Postal Service (USPS) use to do the same for packages. The truck also isn't going to roll in icy conditions, or snow, or maybe even in heavy rain.

By constraining the problem, that is by cutting off all the messy bits that require the adaptability of a human, TuSimple hopes to create what are more or less land trains—a faster and more versatile alternative to railroads, built from trucks.

A fully autonomous truck doesn't take breaks. It doesn't eat or use the toilet or have to obey Department of Transportation rules about how long it can drive before it needs to get some sleep. A typical semi-trailer truck can go 2,000 miles on the 300 gallons of diesel it carries, and it can refuel in under half an hour. This means an autonomous truck could travel from Los Angeles to Florida in two days.

Trucks that never stop rolling would be a game changer for transportation planners in any number of ways. The most expensive way to ship goods is by air freight, and trucks that could roll 24/7 could take a substantial portion of the packages that currently must be transferred to planes. They could allow logistics companies to build larger fulfillment and distribution centers farther from the secondary and tertiary sortation warehouses they feed. They could compete with rail on price and planes on speed.

They would also be eerie, especially at night. Imagine yourself pulling alongside a semi and glancing over, only to discover the truck you're passing has no driver—a ghost truck gently twisting its own wheel. Eventually, autonomous trucks would probably eliminate the cab altogether and would resemble the vehicles being pioneered by Volvo, in its Vera program. These "trucks" are just sleds not much taller than their wheels. A trailer sits on their back half, while the front extends forward, so low to the ground you might miss it if you were riding alongside in a conventional truck. Without their trailers, Volvo's Vera trucks look like oversize sports cars, only there's nowhere for a person to sit. Their entire volume is filled with computers, sensors, batteries, and electric motors.

TuSimple is already hauling mail for the USPS from Phoenix to Tucscon. In a pilot project, TuSimple hauled mail from Phoenix all the way to Dallas, a run that takes twenty-two hours to complete. Normally, that requires either multiple days—at least two but possibly more—or a team of drivers who sleep in shifts. A fully autonomous truck on the same route wouldn't have to stop at all. Packages dispatched by 8:00 a.m. Sunday would arrive by 6:00 on Monday morning, something impossible to achieve today without team drivers or the use of an airplane.

Autonomous trucks would not be the end of truck drivers like Brett. But, as is often the case with automation, they might eliminate many of the tasks that Brett is currently saddled with. In place of traditional long-haul trucking, with all its grinding exhaustion and time away from home, there might be a need for more drivers who ferry goods on shorter routes, through environments more challenging than long stretches of barely differentiated interstate highway.

People might still be needed to drive trailers from a distribution hub to a store, say, or from a fulfillment center to a sortation center. If e-commerce continues to eat the in-person kind and we all find ourselves going to the store less often, self-driving trucks could make moving goods cheaper and easier in ways that both enable and encourage that trend. The overall effect could be the displacement of some truckers, but more opportunities for them in different but related roles.

Some see this as an opportunity to end the wage slavery of long-haul truckers, the relationship- and soul-destroying nature of their work, the alienation, the exploitation of people by companies and systems that pay them by the mile in a form of twenty-first-century piecework not so different from the kind that still burdens many garment workers.

For all the TED Talk–style hand-waving and hypothesizing by analysts, technologists, and think tanks, it remains to be seen what the real-world impacts of autonomous trucking will be. A world in which autonomous trucks have taken all the plum gigs—regular routes over long distances, for example—is just as likely to be one in which independent contractors continue to be exploited as they haul goods between the hubs served by autonomous trucks, says sociologist Steve Viscelli.

In this future, something like an "Uber for freight hauling" would mean the same downward pressure on wages that such marketplaces for labor have always meant. As one economist who worked with actual Uber data discovered, the wage that workers in such fluid, two-sided markets for unskilled labor inevitably garner is whatever the local minimum wage happens to be.

It would be a future for truck drivers not unlike our present for workers in, for example, Amazon warehouses. It would mean more jobs, but all of them more surveilled, controlled by algorithm, and Taylorized than ever.

Chapter 14

WHAT ACTUALLY HAPPENS INSIDE AMAZON'S WAREHOUSES

←

There's a reason Amazon is so eager to give people tours of its massive distribution centers. Hulking and featureless on the outside, they are, like some kind of asteroid-size geode, a wonder within.

Graft Willy Wonka's sense of whimsy onto Henry Ford's pragmatism, hire M. C. Escher to decorate and Rube Goldberg as chief engineer, then crib the scale of the place from the final scene of *Raiders of the Lost Ark*, in which a warehouse of crates stretches to the vanishing point. Make the ceilings snow white, the floors polished concrete, and fill the guts of the thing—miles of curving stainless-steel conveyor—with tens of thousands of daisy-yellow plastic totes.

Standing on the second-floor mezzanine at one end of the facility, looking down its improbable length, the eye alights on a Fanuc industrial robot arm two stories tall. Also yellow, it swings to and fro.

The arm grabs a yellow bin, a "tote" as they're known within Amazon, from a conveyor. Using its purpose-designed actuator, it lifts its mass with a speed suggesting it weighs nothing at all. It pauses at the top of the arc of its movement, as if taking a moment to think. Then it places the tote in exactly the right position on a growing, three-dimensional Jenga of identical totes atop one of a half dozen nearby wooden pallets. The whole sequence takes less than three seconds. Worldwide, such palletizers have stacked more than 2 billion totes.

If you defocus your eyes, shut out the patter of your Amazon-provided, hi-vis-vest-wearing tour guide, and take a moment for yourself at the foot of this silent, unceasing, fully automated palletizing robot, you can use your imagination to pierce the veil between the present and a future in which such robots are everywhere, doing everything, their dexterous tentacles even building copies of themselves in mechanized brood chambers smelling of ozone and machine oil, the room pitch black save the occasional flash of a laser scanner.

"People always ask if we are going to replace humans with robots, but since introducing robots to our warehouses, we've hired over 300,000 associates," your tour guide is likely to say, or some variant of it, because it's what employees of Amazon always say, over and over again, with little variation, in a way that suggests it's written on a sheet of talking points widely circulated within the company.

Like Darwin's Galápagos finches, the more than 200 semi-automated "fulfillment centers" in the United States in which Amazon stores its inventory and distills it into individual orders are all identifiable as members of the same lineage. Yet each of these warehouses is also, on close inspection, different. Since Amazon's operation and robotics teams are constantly learning and incorporating those lessons into newer facilities, it all depends on when each fulfillment center was built.

The latest generation of semiautomated fulfillment centers includes two—or sometimes one—"pick towers." Pick towers are two- and three-story structures in which hundreds of "drive units," Amazon-orange robots resembling oversize Roombas, with eight-foot-tall shelves of goods riding on top of them, scurry about in a caged-in, human-free zone. Worldwide, Amazon has more than 200,000 of these robots. Collectively, they've allowed the company to store 40 percent more goods compared with warehouses without them.

There's so much to explore here, so many layers to unpack, a not insubstantial amount of drama in the lives of the humans who spend ten and sometimes twelve hours a day embedded in this system like bugs in amber. As ever, we'll thread the eye of otherwise incomprehensible complexity by following the path of our humble USB charger. What follows is an account of how goods move through a sort of platonic ideal of a fulfillment center, informed by the accounts of workers at Amazon's Shakopee, Minnesota, fulfillment center just outside Minneapolis, and also by research and reporting at other Amazon fulfillment centers of the latest generation, most notably the one in Baltimore, Maryland.

The technology in each of these fulfillment centers is nearly identical, but the layout changes as the industrial engineers who design these facilities learn from each one they build and operate.

Once our truck has backed into the loading dock, the building and its trailer are temporarily mated into a single contiguous workspace. Amazon associates—what Amazon calls its employees—begin the process of pulling goods out of the truck and into a staging area 150 or so feet deep, running the length of one side of the fulfillment center.

Here, as is often the case at the beginning of a thing, it's all a bit messy. The most sophisticated tool an associate uses is a pallet jack, which looks like the tines of a forklift have been sawed off and then welded onto a moving dolly. Its smooth metal wheels, each wrapped around a little fist of ball bearings that chase each other around their axles, make it appear to float just high enough off the ground to fit into the slots of a wooden shipping pallet.

With the use of the pallet jack, each pallet of goods, some of which can weigh up to a ton, is wheeled into the staging area. Up to this moment, it's very likely that a shipment of goods has come all this way as it would travel to any other retailer, wholesaler, or distributor—in a box whose contents have remained untouched since it left the factory.

If an item is too large or bulky to go into the cubbies that make up the majority of storage in a fulfillment center, and which range in width from six to eighteen inches, it may stay on a stack on its pallet and then be diverted to another part of the fulfillment center. The same area accommodates pallets of homogeneous items that arrive in

large numbers and sell quickly, such as Amazon's various Echo speak-
ers during Prime Day, Amazon's annual shopping holiday.

Everything else, whether it's a stack of identical boxes of the same
product or any of the millions upon millions of varied and loose
packages, gets broken down by humans in the old-fashioned way:
"depalletized" first, if it's on a pallet, and then unpacked using box
cutters and work-gloved hands. The diversity of these goods come
to Amazon from a dizzying array of suppliers, including wholesalers,
manufacturers, and small home-based businesses. Items also come
from bargain-hunting resellers who find deals in rural Walmarts, and
even, in rare cases, fish items out of the trash.

The sheer diversity of these items and their sources is the reason
that a company as technologically savvy as Amazon has added more
humans to its ranks in the past half decade than any other private
employer in the United States. One of the engineers who designed
Amazon's systems calls the flow of items through the company's warehouses "goo" to denote both that the items involved are so numerous
that it's as if they are particles in a liquid, flowing through the supply
chain, and that the varied nature of this goo makes its passage tricky
and the opposite of frictionless. Dealing with this sprawling, inho-
mogeneous mass requires the intelligence, dexterity, and problem-
solving abilities of many, many humans.

The bar code on each of these incoming boxes is scanned by an associate, the first step in its ingestion into the warehouse management
system, a species of cloud-based software that runs Amazon's fulfillment centers. This software will track an object from this moment
all the way to its delivery to a customer. Fast, cloud-based databases
might seem mundane, but it's worth noting that nothing that will
happen next in this object's journey through the warehouse would
be possible were it not for the utter ubiquity of modern information
technology. And the IT backbone of Amazon's retail operation is, of
course, its own industry-dominating cloud services, known as Amazon
Web Services (AWS). AWS encompasses nearly 200 (and counting)
cloud-based IT services, from databases and cloud computing to AI
and general-purpose storage.

Our cardboard box of several dozen individually packaged USB chargers is then processed by a "receiver," a person who inspects them, scans the individual bar code on each item, confirms the items are intact and that they match their descriptions, and then drops them into yellow totes.

These totes, rectangular prisms twenty-two inches wide, twenty-two inches deep, three feet long, and made of durable, strong, recyclable plastic, are to Amazon's individual fulfillment centers—and, increasingly, Amazon's entire distributed network of warehouses—what the shipping container is to global trade. Just as a shipping container can be filled at a factory or warehouse, trucked to a port, transported on a ship, then lifted onto a train or a truck and brought to another warehouse or even its final destination, these yellow totes are the default transactional unit within Amazon's supply chain. They are, like shipping containers, an "intermodal" means of protecting and transporting goods, one that makes their passage markedly easier than trying to move a diverse hodgepodge of individual items.

Another way to put it: just as the internet is a system designed to quickly transmit uniform units of data, known as packets, so too is Amazon's supply chain a system designed to quickly transmit these yellow totes.

Indeed, a primary reason that Amazon hasn't created a system that uses nothing but totes for all transportation and storage, creating an end-to-end infrastructure that copies the architecture of the internet but applies it to physical goods, is that there's too much empty space inside most totes. The company can't get the density it needs within the storage areas in its fulfillment centers, which is where our tote, USB chargers nestled inside, is headed next.

▶ ▶ ▶

Our USB charger will soon be transported on a conveyor, not to be confused with a conveyor belt, and yet also after all the direct descendant of Ford's conveyor belt system. The first conveyor belts were little more than long strips of leather on rollers, used to pull ore out of mines. From Ford's invention on, they became more common but not much more sophisticated. At Ford, ladder-like ramps of steel rollers used gravity to move items through factories. In one of Ford's factories, the smelter was actually placed at the highest point of one end of the factory, so that parts could be carried through the rest of the manufacturing process via gravity.

Later, motors were incorporated, every twenty feet or so, into these systems of rollers, and each roller was connected to the next via small belts. In this way, the motorized conveyor was born.

In the early 2000s, electric motors became small and powerful enough that it occurred to engineers at a company called Intelligrated to put them inside each roller. Making it possible for rollers to move independent of one another allowed for fancy tricks like conveyors that could sort boxes to one side or another at a fork in their path or turn boxes so they are oriented in a particular direction.

Esoterica like this, and the insights into how Amazon's warehouses work, some of which have never been made known outside the company before, largely come by way of the one man who deserves more credit than anyone else for transforming the structure, function, and efficiency of Amazon's fulfillment centers. That he deserves only a plurality of the credit, perhaps barely a double-digit percentage, hardly matters for our story. From the perspective of how Amazon automated itself and may someday automate its competitors out of business, perhaps no one else is more central to this tale or saw more of how it all unfolded.

In April 2002, Mick Mountz began working full-time on the idea that became Kiva Systems, a robotics company that almost couldn't get funded because dozens and dozens of venture capital investors said no. VCs are people whose entire job is to bet on longshots. But robots? For retailers? You've got to be crazy, they told him.

Mick was living off his savings, but he wasn't exactly eating ramen and sleeping under his desk. Previously, he'd been an engineer at Motorola, where he learned how microchips—the most complicated and difficult-to-make objects on Earth—are manufactured. Then he was a product manager at Apple, on the Power Mac team, the Macintoshes that Apple sold from 1994 until 2006.

After Apple, Mick went in an unexpected direction and made a decision that may one day be viewed by future historians as pivotal for the trajectory of e-commerce and the nature of work: Mick went to Webvan.

People who were sentient during the first dot-com bubble of the late 1990s remember the delivery service Kozmo, because you could order literally anything: a can of Coke, a pack of gum, a CD. (Recall that the first iPod wasn't released until 2001.) Whatever it was, Kozmo had to deliver it within thirty minutes. With the benefit of hindsight, it's clear the whole enterprise epitomized a stock market bubble so inflated, an era of optimism so wild-eyed, it was as if investors, especially the ones who were supposed to know better, had truly lost their minds. Webvan, slightly more sane than Kozmo and therefore only a tenth as famous, just delivered groceries, and only in a limited area.

In 1999, at the absolute height of the dot-com madness, Mick was hired onto the "business process team" at Webvan. "They'd already opened their first warehouse in Oakland in the summer of '99," he recalls, "and we were asked to look at next-generation processes to scale this thing to profitability."

Like its competitor Kozmo, Webvan was losing money on every order. The real black holes for cash were fulfillment—getting the orders out of the warehouse—and delivery.

After working on the issue for a year, and seeing the writing on the wall, Mick left in 2000, and Webvan filed for bankruptcy in 2001. But the idea that it might be possible to use his engineering and manufacturing background to solve the problems he had seen inside Webvan stuck with him. Like all inventors, authors, entrepreneurs, outsider artists, and other determined weirdos, he could not shake the feeling that he could be the one to craft something that could disrupt a whole industry.

An ethnographer of the inanimate, Mick traveled the country to examine the devices that were already automating the handling of goods inside warehouses. At the time, the state of the art was the "tilt-tray sorter," a conveyor system on which each segment could be tilted in one direction or another to drop goods into a tote somewhere along its length.

"The tilt-tray-sorter guys would say, 'Use this,'" he recalls. "And I'd say, 'What about a jar of Prego?' They'd say, 'Not a breakable glass thing.' 'Pepsi?' 'Not a round rollable thing.' 'Eggs?' 'No.' Everything had a constraint."

And that was the trick: While Jeff Bezos's "everything store" was still mostly hypothetical—the company sold nothing but books until its first tentative steps toward selling other goods in 1998—Mick was determined to create the infrastructure that could enable it, not specifically for Amazon but for any company that wanted to sell goods online and deliver them quickly. He knew that what he was building had to work for pretty much any consumer item up to a certain size, from office supplies and automotive parts to cosmetics and toys, or else he'd never have enough customers. This being a robotics and hardware company, notoriously capital-intensive fields of endeavor, he'd need as many takers as he could get.

Investors told him he'd need at least $100 million to get to profitability building a company capable of creating the systems he envisioned. As of January 2004, he had $1.6 million. That year, with a handful of engineers, he got started in a small office in Burlington, Massachusetts, a town just outside Boston. His embryonic start-up may have been a continent away from Silicon Valley, but it was well placed to take advantage of the talent pouring out of the many universities in the area, not least Mick's alma mater, MIT.

The fundamental challenge Mick and Kiva Systems faced was the same one that held back all of e-commerce: the world's supply chains were not designed to handle individual items. Rather, they were built for containers, truckloads, pallets, and maybe, if your supply chain manager was particularly savvy, cases of goods. Various pallet- and case-level systems for handling these goods in warehouses were in use even in the late 1990s. Some of them were fairly sophisticated even then, most famously the ones pioneered by Walmart.

Before Kiva, for nearly all retailers, automation in the supply chain ended at the distribution center. Unlike Amazon's modern, automated, e-commerce-focused fulfillment center, the venerable and decidedly not high-tech distribution center was for retailers the last stop for goods before they were sorted and consolidated onto a different set of pallets and then sent by truck to stores. At the store, employees unloaded products into a stockroom and then onto carts and then onto store shelves, in a process fundamentally unchanged since employees began stocking shelves for the opening in 1916 of the world's first self-service grocery store, Piggly Wiggly.

Starting in 1976, Price Club made a virtue of this supply chain by inviting customers into what was essentially a giant stockroom, eliminating the front of the store and all the labor and other costs required to shelve and sell smaller quantities of goods. In 1983, Costco and Sam's Club opened; both were predicated on the same model. Walmart also borrows from this model. If you've ever noticed the shipping pallet under a big box of goods in the middle of a wide aisle in a Walmart, you've experienced a retail experience designed not so much for you as for the distribution center–centric supply chain that put it there.

The reign of big-box stores might have continued indefinitely were it not for the internet. E-commerce hasn't exactly been the asteroid strike to end all lumbering retail dinosaurs, but then again the actual dinosaurs didn't die out as quickly as once thought, either. As of this writing, e-commerce sales in the United States are, by the most commonly cited official measure, about 16 percent of all retail sales, after growing an astonishing 50 percent in the first half of 2020, thanks to the pandemic. By a less commonly cited measure that looks at just the sorts of products that are typically sold online, and excludes sales of things like gasoline and automobiles, the figure is quite a bit larger: at least a quarter of all retail sales.

If, as investor Bill Gross has suggested, timing is the single most important factor in the success or failure of a new tech start-up, Kiva launched at precisely the right moment. Firms that served businesses, especially, were already seeing a need for fast, consistent delivery and had the kinds of relatively price-insensitive customers who would pay for it.

"Our first customer was Staples," says Mick. "They were an e-commerce pioneer. They had a two-day fulfillment network years and years ahead of Amazon."

Staples' first pilot with Kiva was in 2005, the same year Amazon launched Prime, which at the time cost $79 a year for free two-day delivery nationwide.

"Staples had fifteen distribution centers around the country with two-day delivery to 70 percent of desktops in America, meaning they would drop off the package on the office manager's desk," says Mick. In his retelling, Staples signed up for Kiva almost as soon as they heard about it because the company was locked in an existential battle with OfficeMax. (The two attempted to merge in 2015, citing Amazon as the real threat, but were blocked by the Federal Trade Commission.)

"They were both selling the same Post-it notes," says Mick. "The only difference was fulfillment. This was when fulfillment became a competitive weapon."

Kiva's success at Staples quickly led to deals with Walgreens, Zappos, Gap, and Diapers.com.

This did not stop analysts from doubting that Kiva was a long-term solution to anyone's warehousing and fulfillment problems. By 2011, the company had been cash-flow positive for two years and was preparing for an IPO. "As we were talking to Wall Street analysts, they would say, 'If your stuff is so good, how come Amazon isn't using it?'"

What Mick couldn't tell them was that he had been in talks with Jeff Wilke, then Amazon's vice president and general manager of worldwide operations, since the fall of 2003.

Surprising everyone, Amazon acquired Kiva Systems for $775 million in 2012.

Amazon had already acquired Zappos and Diapers.com, both of which used Kiva. Seeing the systems in action is what convinced Amazon's leaders they needed to have Kiva all to themselves.

"When I agreed to sell the company to them, we had an understanding that we would be a standalone company, and after two years we could go back to the market and serve our [existing] commercial customers," says Mick. But it was a verbal agreement, and when two years had passed, Amazon's leaders decided against continuing to service potential competitors.

The result was a mad scramble for an alternative to Kiva by every other company doing fulfillment of e-commerce orders, from Amazon's direct competitors like Walmart and Wayfair to the many third-party logistics companies performing those functions for companies that don't want to do it in-house. That "robot arms race" is continuing to this day, as companies struggle to keep up with Amazon's ever-faster pace of delivery. The "Amazon Prime effect" is the reason why so much of what happens in an Amazon fulfillment center is relevant to the entire $5.5 trillion supply chain and logistics industry—because in a way, it's the future of all of it.

That Amazon, with the help of Kiva Systems, has created an entirely new and different sort of supply chain is the heart of the difference between Amazon and its closest competitor in retail, Walmart. It's the reason Walmart is the largest private employer in America, with about three times as many stores in the United States as Amazon has facilities of any kind in the entire world, yet Amazon is worth four times as much as Walmart and Jeff Bezos is one of the richest people on the planet.

At the core of the difference between Walmart and Amazon is that Walmart's logistics network is still for the most part based on the distribution center, which is a warehouse designed to serve stores and handles pallets and cases. It's the kind of facility all retail chains, and the wholesalers who supply smaller independent stores, have relied upon almost exclusively until very recently. Amazon, by contrast, uses fulfillment centers, which e-commerce companies had to improvise as they tried to figure out how to ship individual items to customers without losing money on every transaction, as Kozmo, Webvan, and so many other dot-com flameouts did.

The roots of Amazon's fulfillment centers were, not surprisingly, in Walmart's distribution centers. In the early years of Amazon's explosive growth, CEO Jeff Bezos hired Jimmy Wright, formerly a key executive in charge of logistics at Walmart, to build out Amazon's network of warehouses. Wright lasted a year before deciding to retire; his successor was Jeff Wilke, an engineer, MBA, and logistics expert who was then working in the pharmaceutical business of manufacturing giant AlliedSignal, according to Brad Stone's definitive history of the early years of Amazon, *The Everything Store*.

The chaos that resulted from trying to run Amazon's supply chain, which had to get individual items to end consumers in the same way Walmart got pallets of goods to stores, inspired Wilke to borrow directly from his background in making things at scale. He looked at Amazon's warehouses and saw not distribution centers but factories.

"We were essentially assembling and fulfilling customer orders," Wilke told Stone. "The factory physics were a lot closer to manufacturing and assembly than retail." Wilke dubbed these reinvented warehouses "fulfillment centers."

The purpose of every e-commerce fulfillment center on Earth is to efficiently deliver the "each pick"—storing, boxing, and shipping out individual items. What follows is how this process works in an Amazon fulfillment center, but in its fundamentals it is remarkably similar across a broad array of Amazon's competitors. I can say that with confidence because, whether they were delivering groceries to London and its suburbs, tools and spare parts to the whole of the United States, or a grab bag of raw materials and household gadgets to and from ports in Vietnam, I've spent a substantial amount of time inside such facilities, interviewing those who built and work in them. I've also researched and reported on an even greater variety of such facilities.

Many of the core principles at the heart of this system are surprisingly universal and have apparently been arrived at independently by a number of firms and vendors. These systems are, like the dawn of Taylorism, the advent of the shipping container, and the perfection of mass production, another instance of multiple discovery, of smart and determined people arriving at a common solution because it is, given current technology and the limits of human imagination, the best we can do.

▶ ▶ ▶

Back to that tote full of USB chargers, including the one we've been following all these many miles. On a pallet jack, the tote is now transported to the station of an Amazon "stower," where it will then be put on a short conveyor, which holds each stower's queue of totes to be processed.

Today, in a fulfillment center in Shakopee, Minnesota, that stower is Tyler Hamilton. Tyler is twenty-two years old, tall, pale, and wiry. I first met him on a cold November day in the basement of the Bethany Lutheran Church in Minneapolis at a meeting of the Awood Center, a community organization whose mission is to "build economic and political power amongst workers in the East African community of Minnesota."

Tyler isn't East African, but he has walked out of his job in solidarity with his largely Somali coworkers, in order to protest working conditions at the fulfillment center, where most of them work.

The night we meet is an auspicious one for such a small community organization. In addition to representatives from the Teamsters and other unions, the room includes a half dozen reporters. On that month's cover of *Wired* magazine is Nimo Omar, a community organizer at Awood, her face framed by a hijab, her Mona Lisa–esque smile suggesting both pride and defiance. Ostensibly, we're all there to hear about the launch of a new, nationwide coalition of community organizations that aim to pressure Amazon and organize its workers across the country.

But I'm there mostly because I figure one way to get some unvarnished truth about what it's like to work inside an Amazon fulfillment center is to talk to people who continue working for Amazon but are willing to speak out about how they wish their jobs were better, whatever the consequences. It is, I am hoping, a way to find some middle ground of real talk between the presumably handpicked associates Amazon makes available to the press through official channels and those who have burned out on the job and are embittered by their employment at the company. (Eventually, I'll end up talking to people who fit those descriptions as well.)

Talking to Tyler, it becomes clear that his job is incredibly physically demanding, but he is good at it. He often beats by 20 to 50 percent the infamous "rate" that governs nearly all entry-level jobs in Amazon facilities. That means, when stowing, grabbing as many as 400 items an hour from totes and stuffing them into bins on mobile shelving, over and over again, for twelve hours a day. He gets thirty minutes for lunch and two more short breaks per shift. Every once in a while, his managers give him a shift that lasts almost but not quite twelve hours so they don't have to give him a third break.

Other than those periods of rest, he is not allowed to sit down. He can take additional breaks to use the bathroom, but if he lingers too long, the system that rules his work life will nag him for excessive "time off task."

Tyler approaches his job like a professional athlete and has built his life around a routine of work, rest, and recovery, buttressed by healthy eating. He says it allows him to do his work in a manner that's sustainable, at least for now, at his age.

Tyler works the night shift, from 5:30 p.m. to 6:00 a.m., three days a week, Thursday through Saturday, instead of the four days of ten-hour shifts that are typical for most Amazon warehouse workers during nonpeak times. During peak, in the run-up to Christmas and on Amazon's Prime Day, workers must be available for mandatory overtime that takes their shifts up to twelve hours.

"Usually I'll lose most of Sunday for recovery," says Tyler. "I try to wake up in normal time to eat something and drink something. Because if you're sleeping after wearing yourself down and you don't do that, your body runs out of building blocks to rebuild, and you'll start to feel shitty for sleeping too long. And then your body will start pulling building blocks from elsewhere. So I wake up in normal time even if I'm tired, I have a drink, I hydrate a lot. I eat something, you know, some eggs and some good stuff. Then I go back to sleep, have a nap. Naps are so wonderful. You don't appreciate them when you're younger."

The reason Tyler—a perfectly healthy twentysomething of what appears to be above-average athleticism and general health, especially for an American—has to do all this is that while his boss is technically a human being known as an "area manager," in reality his life is ruled by an algorithm.

To understand what that algorithm "wants," why it was designed the way it was, how that impacts Tyler's life and the lives of more than a million other Amazon workers across the globe—plus the millions of additional people who are or will someday live under its yoke as all of retail, a $25 trillion industry, is forced to change how it functions in order to either keep pace with Amazon or perish—we have to dive into the system of which it's a part. We have to understand the what, how, and why of the machine that is Amazon.

Stow is the point at which Tyler interfaces directly with that machine. It's also the point at which our USB charger is ingested into Amazon's most ingenious, complicated, and automated mechanical system—the one invented by Mick Mountz and his team at Kiva.

At the heart of this system is a principle borrowed from the design of computers, and specifically the way they store data.

In the early days of computing, data, represented by 1s and 0s, had to be stored on punch cards, magnetic tape, hard drives, CD-ROMs, or a dozen other formats that all had this in common: the speed at which they could offer up data to the microchips that needed it was limited. To get to a particular chunk of data, the tape had to be advanced to the right point, or the arm that reads tracks on a spinning hard disk had to move to a certain point in space.

Various forms of "volatile" memory—the sort that disappears as soon as you cut the power—had long existed, and they offered the advantage of fast access to any of the data they stored solely through the electrical switching of transistors. The best known kind of volatile memory is RAM, an acronym for random-access memory.

But even RAM wasn't enough for successive generations of ever faster, ever-data-hungrier microchips. From 1980 on, CPUs (central processing units, the primary "brain" of any computer) became faster at a rate that outpaced the rate at which their memory, that is RAM, could feed them data.

To get data to these chips as quickly as possible, another layer of volatile memory was embedded very close to or even within the chips themselves, and this memory is called a "cache." Caches are now so important to modern CPUs, which run every computer you encounter on a daily basis, from your phone and laptop to your car and countless invisible, embedded computers in household gadgets and connected civic infrastructure, that under a microscope, the physical area of a CPU taken up by these caches can be greater than half its total size.

Having helped to design systems to manufacture microchips, and having run a division of Apple Computer, Mick understood the basics of CPU architecture. He could see how it could be applied to storing physical objects, rather than bits.

Thus was born "random stow," or the idea that the best way to get goods into and out of shelves in a warehouse is to toss them anywhere they'll fit, rather than worrying about storing them according to some sort of organizational system. It worked because, as with any system based on the principle of random access, the stuff that's close enough at hand, that's in the cache of a CPU or a warehouse, is all pretty much equally and quickly accessible.

When a yellow tote full of goods rolls on a conveyor to the work area before Tyler, he grabs the item indicated on the display above his station. What happens next is pretty close to what you do when you use the self-checkout lane at a grocery store.

First, he scans the bar code of the item on either a stationary scanner above his work area or using his wireless, battery-powered hand scanner. Then, instead of putting it into a grocery bag, he stuffs it into one of six- to eighteen-inch-wide bins on the mobile shelving unit before him. Until recently, an associate like Tyler would then have to scan a bar code on the bin he just put the item in so the system knows where he put it. Some Amazon fulfillment centers still work this way. But in its eternal Taylorism- and Fordism-driven quest to excise every possible micron of wasted time and effort from its systems, Amazon has figured out a way to eliminate this step at its newest fulfillment centers.

Amazon now has a system, called NIKE, that includes a camera that can see nearly all of Tyler's work area and watches his every move. It uses artificial intelligence to determine exactly which bin he put an item into. Systems like these don't come cheap, in terms of research and development, and must be trained on thousands, even millions of hours of video, much of it annotated by actual humans in order to create suitable training data. But for Amazon, a technology company, throwing a significant amount of computing power and the expertise of some of the best-compensated engineers on the planet at the problem of saving its hundreds of thousands of human warehouse associates a fraction of a second in a sequence they perform hundreds of times an hour makes perfect sense.

The system feeding Tyler items to store doesn't care which bin he puts them in. "We don't tell associates where to put inventory on the shelf," says Brad Porter, who was vice president of robotics at Amazon until he left in August 2020. "They put it in whatever bin they want, which means they don't have to spend time waiting for the computer to tell them which pod to put it in." ("Pods" are Amazonese for the combination of shelves and robots in a warehouse, while bins are the cubbies in a pod.)

Not overmanaging where people stick things on any one mobile shelf also helps keep items flowing onto those shelves as quickly as possible. "We receive a ton of inventory every day," adds Brad. "So one of the process-efficiency pieces is that if you're trying to put a lot of structure in how you store everything, you create bottlenecks in the flow."

The system does care which mobile shelving unit items go into, however. That's why it's not up to Tyler which items have been delivered to him in which totes, and which bin on which pod he'll be asked to stuff them in—all of that is decided by algorithms running under the umbrella of Amazon's warehouse management system, and more broadly, its globe-spanning inventory management system.

By asking Tyler to put our USB charger in one bin, allowing him to stick an identical one in another bin of his choosing, and, overall, distributing the dozens of chargers that arrived at this distribution center in a single box across dozens of mobile shelving units, the random stow system maximizes the chances that a mobile shelf with one of these chargers in it is readily accessible when an Amazon customer orders one and it must be extracted from the storage system and delivered to that person.

In order to make rate, every fourteen seconds or so, Tyler must grab an item, scan it, and stuff it into a bin.

Chapter 15

THE UNBEARABLE COMPLEXITY OF ROBOTIC WAREHOUSING

It's at this point in the story that a reasonable person asks why, in a company that has deployed so many robots, so many humans must perform the many Sisyphean tasks at the heart of Amazon's sorting of goods in its warehouses. The world's first industrial robot was installed in a factory owned by General Motors in 1961; the underlying technology was patented in 1954. Why have we not by now created robotic arms dexterous enough to take over a task as elementary as picking up an object and placing it in a soft-sided cubby?

The short answer to that question is that this task only seems elementary because humans are so incredibly good at it. The longer answer is that this is in fact a task so difficult that, alongside language, intuition, higher-order cognition, and everything else that humans are better at than any other organism in the known universe, achieving it should rank as one of our most astonishing accomplishments.

It took evolution almost 4 billion years, from the first self-replicating molecules to the advent of modern humans, to stumble into a system of manipulative digits of sufficient precision and flexibility—two things any good roboticist will tell you are diametrically opposed—attached to a brain with enough capacity for planning and spatial reasoning, to do what Amazon stowers, pickers, and packers do hundreds of times an hour without even having to think about it. The things we can do with our hands—pick a flower, brandish a weapon, cradle an infant— are an unqualified miracle.

In the early days of pitching the Kiva system, people would ask Mick Mountz why he didn't just use a robotic arm in the stowing and picking stations. "And the answer was, you'd need a NASA-size re- search budget to come up with a robot that could do what the human was doing," he says.

This is not to say that many, many companies, including Amazon itself, aren't trying to replace humans in these processes. In chap- ter 19, I dive into their efforts, their limitations, and what they can tell us about the future of all manual labor, from manufacturing and logistics to the service industry. But the thing to know about all predictions about an imminent future of "lights-out" warehouses in which there are no humans at all is that no one knows when or even if that future will arrive. It's certainly not going to be anytime soon.

The problem of reproducing the dexterity of the human hand and the capabilities of the nervous system directing it even has a name among engineers and academics: Moravec's paradox.

Hans Moravec is an adjunct professor at the Robotics Institute at Carnegie Mellon University. Here's how he described the observation that bears his name in his 1988 book, *Mind Children: The Future of Robot and Human Intelligence*: "It is comparatively easy to make com- puters exhibit adult-level performance on intelligence tests or playing checkers, and difficult or impossible to give them the skills of a one- year-old when it comes to perception and mobility."

One formulation of Moravec's paradox goes like this: it's far harder to teach a computer to pick up and move a chess piece like its human opponent than it is to teach it to beat that human at chess.

After making that remarkably cogent observation, Moravec goes on in this and subsequent books to claim that the gulf between the intelligence and dexterity of computers is so profound that while human-level dexterity is nowhere in sight, we'll soon have human-level intelligence in computers. Of course, the idea that we're anywhere close to that sort of intelligence in machines has become a shibboleth of its own, one denoting a species of since-discredited optimism about both how quickly our computers would evolve and how much we thought we knew about how intelligence works in the first place. (A good general rule when listening to any expert: their skepticism is often on the mark, their optimism frequently way off.)

When we talk about the capacities of computers and artificial intelligence, the measure of their abilities is most often something like the Turing test, which asks whether a computer can convince humans chatting with it that it is in fact human. But I would argue that Moravec's test, which I imagine as his paradox reframed as a test of robots' abilities, would be a better measure of the capabilities of our synthetic offspring. Humans, after all, are surprisingly easy to fool in conversation. Some of the entrants in the annual Turing contest have already duped some of its judges. But whether a machine can physically perform a task as we do, in the real world, is a far less fuzzy metric of their progress.

Once Tyler Hamilton has stowed enough items in the mobile shelving unit before him, it rolls away at a walking pace. Each of the orange drive units that carry these movable shelves weighs 350 pounds and can carry up to 1,200 pounds. In addition to lifting all that weight, they're surprisingly robust, able to roll over drains in the floor and other imperfections without dumping the shelves they carry because their wheels have independent suspension.

These robots exist in their own world within the fulfillment center, literally. Their domain is limited to a broad field of smooth concrete bounded by metal fencing on all sides—with only a few access gates for maintenance and upkeep associates—plus the stow and pick stations where they offer up their bounty to their human coworkers.

No one is allowed on the field where the robots rule, and entering the area is a fireable offense. To step into the robots' realm is to invite a mauling. Like all industrial robots, these are intended to be fast, strong, and segregated from humans so that they can do their job with maximum efficiency.

The exception to the no-humans rule are associates who are wearing a special vest that includes a beacon that alerts the robots to their presence. One reason humans must enter the robots' domain is that the robots break. Another is that things sometimes fall off the shelves they carry, even though every cubby is secured behind an elastic band that stretches all the way around the shelving unit.

One of the associates who is allowed into the robot's world is Samaria Johnson. She goes by Sam and works at the Baltimore fulfillment center. Her bleached, short-cropped hair and tattoos go hand in hand with a certain physical self-confidence; if she weren't wearing a vest, she could easily pass for a personal trainer. She's on the "amnesty team" that keeps the robots on the robot floor operating smoothly. The system may be automated, but it cannot function at peak capacity without a team of minders flitting about in its innards, solving problems that only smart and cognitively flexible humans can address.

"I just like being the behind-the-scenes person where everybody still knows me," says Sam. "It just feels good when you've got people at a [pick or stow] station and they're like, 'You're on my floor today? I'm going to get high numbers!'"

My interview with Sam takes place under the watchful eye of an Amazon press flack, and like everyone I'm introduced to that day, she has nothing but good things to say about the company. I ask her about the constant stream of news reports about conditions in Amazon warehouses. "I hear people with their complaints and I just think, 'Y'all never worked at a warehouse before,'" she says.

Associates at Amazon who stick around more than a year get to "cross-train" in other roles within the warehouse, and that's how Sam got onto the amnesty team. Working directly with the automation, her job changes quickly, since significant shifts in how Amazon's robots function are always just one over-the-air software upgrade away.

"They do updates on the system to fix what we are struggling with," says Sam. "Every time they update, a lot of people don't like it, but you gotta get used to it."

That rapid pace of change is true of Amazon in general. "For us in robotics, that means we need to build things that are flexible," says Brad Porter, former head of robotics at Amazon. "We have to expect that the network and the processes are going to change, and that we are going to use processes in ways that weren't originally intended."

It's the flexibility of robots, especially modern ones, that makes them so useful. Things that today's robots have that they did not fifteen or twenty years ago include: powerful but energy-sipping digital brains, robust wireless connections to the cloud, energy-dense and fast-charging lithium-ion batteries, powerful and compact electric motors, and a surplus of inexpensive sensors that, with only a software update, can enable new capabilities and new modes of operation as engineers dream them up.

"I did some work retrospectively about how we said we would deploy Kiva systems," says Brad. "If you look today at how we deploy it, it's quite a bit different than that original case. If you went building to building within our network, you'd be surprised that they don't all look identical, and it's because they're built each year as our process changes."

Today's robots, even ones as outwardly simple as the drive units that move around the shelves in an Amazon fulfillment center, represent a sharp break with automation of the past. So-called hard automation—all those miles of conveyors, tilt-tray sorters, and finicky, complicated machines for performing specific tasks in precisely one way—represent huge investments that will require even more expense, downtime, and retrofitting any time their owners want to update them to make a warehouse or factory more efficient. This is one reason manufacturing is such a conservative enterprise: the hard automation and traditional industrial robots long favored by automakers, for example, are not easily updated. Amazon's robots, by contrast, are more like ants in a nest or bees in a hive, their very nature dynamic, their aggregate behavior a function of software.

Roboticist Raffaello D'Andrea, cofounder of Kiva Systems, "liked to call himself the chief algorithm guy because it was all about the algorithms you could invent and apply to the warehouse," says Mick. "Once every piece of inventory could walk and talk and move on its own, you could dream up a million different algorithms to move and store those items. It was such a rich environment for creativity."

One such algorithm keeps the drive units from clustering too much during a lull. In the early days of Kiva's deployments for customers such as Staples and Walgreens, their warehouses would shut down on the weekends. Workers would come back on Monday morning to find that all the drive units were gathered around the recharging stations like ants to a picnic. Fixing that behavior with a "spread-out algorithm" was the work of a summer intern at Kiva.

▶ ▶ ▶

The shelf on which Tyler has stowed our USB charger has a life of its own. It may do any number of things once it glides away from his stow station, but the most likely is that it will try to get out of the way, drifting into the huge and ever-shifting grid of movable shelves on the robot floor.

Our movable shelf is not going far, however, because the essence of speed in an Amazon fulfillment center is keeping things close. Some of this behavior is a natural consequence of the design of the system of which these robots are a part. For example, shelves that are needed frequently tend to stay close to the human stowers and pickers who must access them often. Their actions are orchestrated by the omniscient, hugely powerful warehouse and inventory management systems of Amazon, but to a surprising degree, their movement is self-organized.

The amount of time an item resides in an Amazon fulfillment centers varies, and Amazon does not comment on this metric. But in summer 2018, Amazon implemented an index that includes data like how long inventory belonging to a seller on its Marketplace has been in its warehouses. Amazon began charging fees to sellers that didn't move product fast enough, and it has been slowly ratcheting up the score sellers must have in order to avoid incurring a fee.

Vast as they are, Amazon's warehouses are intended merely to be a sort of local cache of goods intended for next-day delivery. The principles of computer architecture that are inherent in the design of the Kiva robotics systems (which is now the Amazon Robotics system) are fractal: these principles repeat themselves at every scale of the system, from the actions of individual robots, to the behavior of all of the robots in a fulfillment center, to the behavior of Amazon's entire network of warehouses, which are constantly rebalancing inventory between each other as well as within themselves.

Something similar happens in computer chip design. Chips have small local caches of data—known as level 1 caches—immediately accessible to processors and also caches slightly more removed both physically and temporally, including levels 2, 3, and 4, depending on their design. These caches contain copies of data that also exists in main memory (RAM) or on the computer's (spinning or solid-state) hard drive.

That redundancy is key, and this hierarchy of caching translates to how fast data is available to the chip. The same principles are at work in fulfillment centers as well.

"It is similar to memory or storage," says Brad. "If you had multiple copies of a memory or document in our cloud computing system, you're going to get faster retrieval across multiple different nodes, some of which are geographically distributed. It's very similar with our inventory: if you want speed of access, it's great to have [multiples of a single item] spread throughout different pods." Random stow, he adds, "happens to also be great for high-speed random access to inventory."

The Kiva system of shelves sitting atop mobile drive units "was a computer-based architecture," says Mick. "In front of the picker, you have a queue of six to eight pods waiting to be processed, and we called that 'L2 cache.' That's the first place we look for something." By this logic, goods stored farther back in the pelagic depths of the storage system are comparable to a computer's "main memory."

Without these design principles, it would be much more difficult, maybe even impossible, for Amazon and, increasingly, its competitors, to offer fast delivery. They're also essential to the next step in the journey of our USB charger: its extraction from the fulfillment center's storage area and robot floor, known as "pick."

Pick works like this: Someone orders their umpteenth USB charger from Amazon. For the company to fulfill its "customer promise," a phrase you hear often when you hang out with Amazonians, the company has to immediately initiate the process of delivering that charger. Right away, it goes into a virtual queue with the nearly 10 million other orders the company receives on average every day. That's 115 orders a second. In the time it took you to read this paragraph, Amazon received nearly 2,700 orders. Another way to think of it: every second, Amazon fulfills a UPS delivery truck worth of orders.

From the moment your order is received to the moment it goes onto a truck to leave an Amazon fulfillment center, the time elapsed is usually between forty-five minutes and two hours. Figuring out how to make that happen is a triumph of computation, software engineering, and artificial intelligence on a par with getting you fast search results from Google or making sure Mark Zuckerberg gets his thousandth of a cent every time you tap on an ad for a novelty T-shirt on Instagram. It is, in other words, one of the hardest and most complicated problems ever solved by humanity.

The central challenge of figuring out which USB charger in which bin in which shelf atop which drive unit in which fulfillment center should be yours is that this problem includes predicting the behavior of dozens or even hundreds of people, machines, and systems whose behavior is, to one degree or another, random. What's the weather like today—and tomorrow? How's traffic? Are people ordering more of one sort of item than usual because Gwyneth Paltrow promoted it in her newsletter? Is there a pandemic causing more people to order online instead of going to the store?

The list of these variables is potentially endless. And yet the algorithms deciding how to optimize the path of the USB charger to the consumer must make a decision quickly. They must, in the language of psychology, be satisficers rather than optimizers—choosing a good enough solution rather than the absolute best.

Even in a hypothetical scenario in which we can stop time and run this computation until the heat death of the universe, the best solution for this item must be balanced against the operation of the entire system—every robot and machine in the warehouse can't just drop everything to prioritize a single order.

Any delivery or routing algorithm must also build some flexibility into the travel itinerary of an item, because there's plenty it can't predict about the next twenty-four to thirty-six hours. What if someone in the warehouse suddenly falls ill? A delivery van breaks down? An unexpected accident snarls traffic? People who write these kinds of algorithms talk about the dangers of overoptimizing them, creating delivery schedules that, like connecting flights scheduled too close to arriving ones, can cause a cascade of failures throughout the system if they don't work out.

Before the system considers any of those possibilities, however, it looks at two basic facts in the fulfillment center at that moment: how fast each of the pickers who will take an item out of storage is picking that day and how many items are already in their queue to be picked. The system is constantly sending those movable shelves toward the picker who can get the item extracted from a bin fastest, and it will reroute the shelves if that picker takes a break, starts to fall behind, or someone else shows up late for work and suddenly becomes available to pick.

One of the characteristics of random stow that makes it so good at getting items to pickers quickly is that if an item is popular, the shelves containing it tend to physically cluster closer to the pick stations. Having your most frequently used tools close to hand was a fundamental principle of scientific management and the foundation of the rethinking of everything from Frederick Taylor's optimization of how machinists did their work to Lillian Gilbreth's design for the ideal kitchen.

But with mobile shelving systems, you get this efficiency for "free."

This is not a special characteristic of Amazon's mobile shelving systems, mind you, but a characteristic of *all* random-access storage systems of its kind, be they digital or physical. The first time I heard of the principle was 4,000 miles away, in an even more automated warehouse in a suburb of London owned by the U.K.-based grocery delivery company Ocado. There, totes of groceries descend into an enormous three-dimensional matrix of goods that are retrieved by robots moving high above, hoisting totes with long cables. The layout and mechanism of Ocado's and Amazon's automated warehouses could hardly be more different, and yet both companies had arrived at the same mechanism for organizing them—random stow.

Full-time associates rotate through jobs at fulfillment centers, so it's entirely possible that the person who stowed an item one day could be picking it the next. This is especially true around Amazon's peak seasons—Christmas and Prime Day—when a large number of goods may be stowed in a fulfillment center in the weeks leading up to that shopping holiday and then are quickly extracted as orders flow in. The result is that many associates may be put to work in one role or another, depending on where they are needed.

Today, Tyler is on pick. In this role, the rate he's expected to meet is similar to the one he was expected to hit when he was on stow. It varies, but a typical rate at a fulfillment center is 400 items an hour, 350 at minimum. Fall below that rate for an extended period of time—associates are evaluated on a six-week rolling average and are given multiple warnings and opportunities to retrain, says Amazon— and you can be fired.

All pick and stow workstations sit on the outside of the robot floor, carbuncles of humanity attached to the engine of commerce within. Tyler's pick workstation is almost identical to the stow workstation. There's the place where the movable shelf rolls up to his station, and next to it a conveyor with totes on it.

After the mobile shelf atop a robotic drive unit arrives, an overhead light shines on exactly the bin containing the item Tyler is to pick. This is a newer innovation in fulfillment centers. It used to be that associates had to look at the flat panel display above their workstation for the letter and number identifier of the bin from which they would pick, but that was one more tiny way they were slowed down and made more error-prone, so no more.

On a display at eye level, a picture of the item appears along with its name so Tyler knows which of the items in the illuminated bin is the right one. Because of Amazon's random stow, the contents of a bin could be literally anything. A not atypical mix of goods in a bin would be a used textbook next to a bottle of shampoo cozying up to a dildo. Amazon sells a lot of dildos. Nearly everyone who has ever worked in an Amazon fulfillment center has a story about the moment they realized this.

"America's appetite for sex toys—indicated by the sheer number and variety of dildos and butt plugs passing through Amazon warehouses—is a subject of fascination to many workers," wrote Jessica Bruder, who interviewed dozens of temporary Amazon workers for *Nomadland*, her chronicle of America's surprisingly large population of itinerant temps.

Because Tyler is a triumph of evolution who was protected and nurtured as an infant by an extended network of caring adults who tolerated in him, as is the case with all human babies, a period of neoteny longer than that of any other animal on Earth, he was able to acquire the intelligence and finesse required to reach into the (frequently overstuffed) bin and, based only on a name and the two-dimensional image he just glanced at, grab just the right item. This ostensibly simple act is something no combination of artificial intelligence and robotics can yet accomplish. In theory, he has to do this every twelve seconds, but because there are breaks when one shelf moves away and another rolls up, he's actually got to do it much faster—something on the order of every seven seconds, in bursts.

Once the item is in hand, he examines it to make sure it's not damaged, scans its bar code on a stationary scanner, puts it in a yellow tote, and then pushes a big gray button over the tote and gives it a shove to send it on its way.

This being Amazon—a website plastered with upsells, including items from your wish list, revealing and occasionally disturbing things other people tend to buy when ordering whatever is in your cart, and paid advertisements for possibly related products—the person who ordered this USB charger has also ordered a couple of other items. Amazon's algorithm has located all of them in a single fulfillment center and knows it can squeeze all of them into one box. Consolidating smaller items into as few deliveries as possible, in the smallest box that will accommodate them, has been of late and probably always will be a priority at Amazon. The company likes to tout this as an effort to be environmentally friendly, but taking these steps is also a way for it to minimize shipping costs.

Now the challenge is getting all three items, which are in different totes coming from different pickers, into a single tote.

This is the point at which the Amazon Fulfillment Engine (AFE) comes in. The company doesn't talk publicly about AFE. Like many of Amazon's processes, the way AFE operates is protected by the kinds of nondisclosure agreements that make former employees shy away from describing it. Luckily, an outbound operations manager who was at Amazon from 2016 until early 2018 documented its inner workings in material intended to educate new associates, which he left on the open web.

"Amazon Fulfillment Engine is a semi-automated sortation system used to process multiple-item shipments (multis)," says the document intended to educate associates and managers about how AFE works. Translation: after items are picked, AFE is the system in which they are all brought together just before being put in a box to be sent to the customer. AFE is "semi-automated" because it is a tangle of hard automation (conveyors, pick-to-light totes, and the like) held together by humans picking up and moving things as only humans can. It's a Mouse Trap game–style agglomeration of conveyors, T-junctions, "re-bin stations," and, for the troublesome orders, a "Problem Solve" area. There's even a zone called "Jackpot." (The documentation I found doesn't elaborate on what you win if your order ends up there.)

The entire AFE process is about circulating and sorting totes on conveyors so that all the totes containing items intended for a particular order end up at a single rebin station at the same time. At the rebin station, a human will take each item out of the tray in which it was riding solo and put it into a single tote before sending it on its way.

It's a perilous journey to that rebin station, however, and revealing of how much process improvement Amazon might squeeze from its fulfillment centers in the future.

First, our tote has to ride down a spiral slide two stories tall. These things look as whimsical as they sound. It's impossible to see them and not want to ride on them, but on my in-person visit to an Amazon fulfillment center, there are signs everywhere warning us not to stick any of our limbs into the warehouse's many conveyors. One Amazon associate told me he heard that someone did ride on a warehouse conveyor, once, on his last day, but I heard so many crazy stories from associates that are impossible to corroborate that I mostly chalk them up to the kind of playground mythmaking that happens any time you confine people in a place like this for long enough.

Let's pretend we can shrink ourselves down to the size of our USB order and ride along, anyway, because there is no other way we're going to make it through the complexity to come without nodding off. Miniaturized and along for the ride in the big yellow tote, our overwhelming emotion would probably be *"Wheee!"* because what comes next happens pretty fast and involves some significant g-forces.

After the spiral takes us down to the ground floor, a "routing sorter" uses software to decide which of the available pathways (i.e., interconnected chains of chutes and conveyors) a tote could journey down actually has capacity for it. Whichever conveyor a tote is directed onto, it then comes to a T-shaped intersection where a human or automated system sends it to the left or the right. Then a human must take the item out of the tote and put it into a tray, which is either eighteen by twenty-four inches or eighteen by eighteen inches. This is the "induction" into the closed loop of the conveyor-and-sensor-and-software-based logic of the Amazon Fulfillment Engine. Like an adding machine of the predigital era, the AFE performs operations using physical systems as much as electronic ones.

Now our USB charger is in the "routing sorter." Our tray is next sent to the "zipper merge," also known as the "induct merge" or "3:1 merge" because technical jargon tends to multiply in closed communities of specialists who only talk to one another. A zipper merge can handle around 2,700 trays an hour, or about one every 1.3 seconds. Some back-of-the-envelope math suggests the trays have slowed to about walking speed at this point. (Elsewhere in the fulfillment center, you can see totes zipping along at ten or more miles an hour.)

The zipper puts two to four trays with different objects on a single conveyor, which you might think is the last step before they arrive before the human known as a "rebinner," but we're not even close to done yet. This conveyor feeds the "main merge," aka the "7:1" or "9:1" merge. Most trays on this conveyor pass to the next stage, but some are recirculated for reasons that are unclear but seem to be related to capacity limitations of this system, which is "the most constrained mechanical system in AFE and, if all is running perfectly, it will dictate your max volume," wrote the outbound manager.

Now our tote goes into the swirling miasma of the "high-speed sorter," also known as the "AFE sorter," not to be confused with the routing sorter it was directed through earlier in this process. If we were still in the bin, this is the point at which our ride becomes less Splash Mountain and more scary boat ride from the Gene Wilder version of *Willy Wonka & the Chocolate Factory*.

Just when we think it's over and we're about to be spat out by this shiny steel whirligig, oops, we're "recirculated" through it again on account of the lane to our rebinner being too full with other trays. Finally, the lane is clear and the tray with our item in it shoots into a discharge lane. At this point, the tray is moving quite fast, and like a spacecraft reentering Earth's atmosphere, it must use friction to slow itself down. To help it, the discharge lane is curved, allowing the tray to shed its extra velocity before it arrives at a rebin station.

Discharge lanes have frequent jams, and when they occur, they muck up the entire Amazon Fulfillment Engine. Engineers have installed cameras that watch these lanes and the entire system and shut it down when a jam occurs. This way, a hapless Amazon associate isn't forced to start scooping up armfuls of books, toiletries, consumer electronics, and sex toys like a latter-day version of that scene from *I Love Lucy* when Lucy gets a job at the chocolate factory.

The entire process from the moment a human took our USB charger out of a tote until now has taken about five minutes, as long as some error such as an unscannable bar code or a clog in a conveyor didn't waylay it.

Now a human grabs the item out of each tray, turns, and drops it into a tote. The trays in which items arrived are then recirculated back into the high-speed engine for getting goods into a particular order and sending them to a particular person that is the AFE.

One wrinkle of how the AFE works is that, owing to limitations on the amount of area it can take up in a fulfillment center and how complex it can grow before the jams that plague it become even more frequent, the entire system only has about a one-minute supply of trays in it.

This means that the humans getting goods into one end of the AFE at the induction stations must be working at almost exactly the same pace as the humans getting goods out of the AFE at the rebin stations, or else one or the other becomes a bottleneck. Both groups must work very quickly in order to keep this all-important system humming. Other upstream factors, like whether pickers are pulling items out of the storage area too fast or not fast enough, and downstream factors, like whether packers putting rebinned orders into boxes are working quickly enough, also affect their work.

Amazon does not give tours of this part of their fulfillment centers. Here's how one former Amazon associate described working in this part of the warehouse in a blog post in 2016: "AFE was a real culture shock. I have worked as a laborer most of my life, but I have never experienced a work environment where employees are pitted against each other as they were in AFE in 2014. There have been some ameliorative changes made through 2015, but the foundational system of AFE remains brutal."

Before Amazon acquired Kiva Systems, AFE was the only method for automatically sorting goods Amazon had in its warehouses, says Mick. This system was designed by Dave Clark, former head of logistics at Amazon and now senior vice president of operations. Even now that Amazon has Kiva for its storage systems, the company still uses AFE, despite its complexity. It's an example of how Amazon has had to balance hypergrowth with efficiency. At the heart of Amazon's past decade is a tension between a need to quickly build new infrastructure like fulfillment centers and a need to make them faster, more efficient, and less costly to operate.

Now in the last tote it will ride in, the USB charger is nestled between the two other items in this order, a pair of objects purchased probably because humans are impulsive and convenience at the level that Amazon affords makes us stupid. Perhaps they're an Arizona Diamondbacks koozie and a mauve women's T-shirt that says MAY CONTAIN WINE. Maybe it's a six-pack of travel-size flushable wipes and a men's hoodie in black that says I SPEAK FLUENT SARCASM. Could be a used copy of *How to Get a Man Without Getting Played* in "acceptable" condition and a novelty dog collar.

The tote containing our USB charger and two other items rolls down yet another conveyor, this one a double-decker so it can accommodate more totes, to a packing station.

On a sunny afternoon at Amazon's BWI2 fulfillment center in Baltimore, I was the packer at one of those stations. Cliff, a packer at Amazon and lifelong Baltimorean, was my tutor. My apprenticeship lasted barely fifteen minutes, but that's about all it takes to learn the job.

Packing stations in Amazon fulfillment centers are laid out one after the other along the conveyor, on both sides, with packers just a few feet from each other on either side of the stream of incoming totes. This is one of the few positions in the fulfillment center that might allow an ongoing, genial patter between workers, the sort that has helped ease the psychological burdens of menial laborers since the days of the pharaoh. But packers generally don't talk to one another, says Cliff. There's too much to do, too quickly.

At each packer's stand-up workstation, there is a shelf of boxes directly ahead, a shiny steel work table at elbow height, and tape dispensers to one side. Based on the dimensions of an item or items, an algorithm has already suggested which size box or padded envelope they should go in. Cliff grabs a cardboard box off the shelf, its sides printed with Amazon's trademark smirk. I assemble it, getting flashbacks to every flat-pack cardboard box I've ever put together when moving house.

A machine spits out a piece of tape of just the right length, and I use it to seal the bottom of the box. Then I, an anonymous packer, put in items destined for an anonymous customer, along with some packing material. These days, the packing material is usually plastic pouches filled with air, about as dematerialized as packing material can get. Then I tape the top of the box, apply a sticker with a bar code on it spat out by a different machine, and drop it on yet another conveyor. Our fully Taylorized, maximally efficient packing sequence is complete, all waste expunged from our every action, the entire process as quick and effortless as it could possibly get, short of the replacement of Cliff and thousands like him by robots.

I ask Cliff if his job is hard. He tells me that it keeps him active, and I believe it. He is well into middle age, his temples and beard graying, but he looks the picture of health. He points to the shelf of boxes at his station and notes that he used to have to reach down to get one, but improvements in the layout of his workstation have made his job easier than ever. The Amazon PR minder standing nearby nods approvingly.

From "pack," boxes go to a device as remarkable as the giant Fanuc robot arm that elsewhere in the facility builds pallets of totes. This robot is called SLAM—scan, label, apply, manifest.

What both the robot arm and the SLAM system have in common is that the objects they are working with are single, discrete, rectangular solids—boxes. As at ports, where automated cranes and guided vehicles for lifting and maneuvering shipping containers are taking over, getting something into a box makes it robot-friendly. Boxes simplify the messy physical world that a robot vision system must perceive and automated manipulators must interact with.

Using a red laser, the SLAM system scans the bar code on the box containing our USB charger and its companions as it rolls past. That bar code links this box to the customer and their address in Amazon's vast databases. A steel and aluminum proboscis then drops onto the box and applies a shipping label. The box is finally on its way to its last destination within the fulfillment center.

This box is now its own intermodal container. Unlike the steel shipping container its contents came so far in, this one is made of cardboard. It travels on conveyors by itself, rather than in a bin or tote or robot shelf. One such conveyor gets it closer to the loading area for the truck onto which it will go. It descends one last spiral chute, then rolls onto its very last conveyor, one different in form from all the ones it has ridden on so far. This conveyor belt—it's an actual belt, for once, like the kind at a grocery store checkout—extends all the way into the back of the truck that is being loaded, bringing boxes to within a couple of feet of the last person to touch them in this warehouse.

A truck-loading associate is one of the last semiskilled blue-collar tradespeople left in the entire warehouse. Deep in the trailer of a truck, they practice an art well known to professional movers: building, one row after another, a stable wall of boxes. It's a challenge of spatial reasoning, like Tetris or medieval stonemasonry.

Big boxes that can stand up to the load of smaller ones go on the bottom. Cracks between boxes must be mentally reserved and later filled with the smallest packages. To make each row of boxes within a truck complete, associates must stretch to get items to the very top of their reach. Height and long arms help. None of this is automated in an Amazon fulfillment center, yet.

Loading docks like this one are almost always hot in the summer and cold in the winter. No amount of heating or air-conditioning can compensate for the giant, gaping hole in the wall where a truck must mate its trailer to the warehouse, and trailers themselves are poorly insulated. While other labor within the warehouse strains workers because of its repetitive nature, these jobs are tough in the same way that all warehouse labor has been since ancient Egyptians built the first warehouses to store grain. Lifting, carrying, moving, and balancing are the whole of the job.

Our USB charger and its companion items are nestled with their packing material inside a delivery box stamped with both the fulfillment center they came from and a bar code telling UPS, the U.S. Postal Service, or the third-party delivery companies that make up Amazon's own delivery service the ultimate destination of this delivery.

Finally, the last box is slotted into the last available space, the rolling gate at the back of the truck is closed, and the semitruck pulls away. The Amazon-branded trailer it's towing is carrying our item to a secondary hub for directing goods to their final destination, known as a sortation center.

It's been two weeks since this USB charger first arrived at the loading dock for incoming items at this fulfillment center in Minnesota. The time a product spends in an Amazon fulfillment center varies greatly, and this is a fast turnaround time for an item. It can take days to process incoming pallets of goods, break them down, and get them into the robot shelves. Waiting just over a week on those shelves before being called up means this is probably a popular item with a high sales ranking.

It's March 20, and the number of cases of coronavirus in the United States is approaching 20,000. In just under a week, the nation will lead the world in confirmed cases. On this day, the governor of Illinois declares a stay-at-home order, and Dr. Anthony Fauci says it's "very difficult to predict" how long such orders will last.

In many places in the United States, panic buying of necessities and staples has been underway for more than a week. As the shelves of retail stores emptied, buyers turned, primarily, to Amazon, which on March 12 declared that all its employees who could work from home should do so. This did not include workers in its warehouses, of course, who were working overtime to meet surging demand. By this point, the e-commerce giant was already sold out of toilet paper and dozens of other household goods.

A study published by a pair of economists in January 2021 will find that all over the world, panic buying reached its highest intensity in the week after governments declared restrictions on movement within a state or country.

It turns out this USB charger, manufactured when the global

pandemic had yet to start, put on a ship just as China began its shut-downs, ordered from Amazon just today, wasn't an idle purchase. It was one tiny element of the mass shift in the global economy from services to goods, from being out in the world to working, entertaining, and learning from home. It's just a tiny, inexpensive gadget, but it's also an embodiment of the journey of billions of dollars' worth of other goods, all part of a tsunami of creative destruction that has permanently transformed the way we live.

Chapter 16

BEZOSISM

←───

For Austin Morreale, working as a stower at Amazon was one of the toughest jobs he's ever loved. The hours were long and the work grueling. The night shift he took at Amazon on top of his day job as a nonprofit case manager was plainly unsustainable, but he had only planned to do it for a summer, anyway. He needed the money, the immediate access to health insurance, and the change of pace. He lasted six weeks.

"I think I'm in a vast minority of people who had an actually good experience there," he recalls, laughing. At the LGA9 fulfillment center in Edison, New Jersey, "it was ten hours of pretty much mind-numbingly boring work, pretty much standing in the same position for the whole shift. But at the end of the shift, I was drenched in sweat and aching like I hadn't ached since I was playing competitive soccer." Austin is fifty years old now; his soccer career ended when he graduated from high school.

"It was physically demanding work," he continues, "but I thoroughly enjoyed it, as crazy as that sounds."

Austin was slow. He kept messing up the patterns for efficient stowing he had been taught. He couldn't "make rate," Amazonese for keeping up with the pace of work. But his managers were generous and "super invested" in helping everyone on his team improve. "My supervisory staff there was absolutely amazing. I've worked for and with a lot of people, and trained a lot of people, but the staff I worked with there are probably some of the best professional folks I've worked with."

On the job, no one ever stood behind Austin and barked at him to work faster. They didn't have to. Twice a day at a stand-up meeting, his shift managers told the group how everyone was doing. "Those numbers are always in the back of your head somewhere," he says.

One day, Austin pushed himself too hard. Light-headed and clammy, he sank to his knees, a no-no at Amazon. Associates are not allowed to sit down while on the job, unless it's lunch time or one of their two fifteen-minute breaks, both of which occur at times designated by management.

"I don't know if it was overexertion or what it was," he says. "My supervisors never themselves made me feel pressure. I put that pressure on myself—'Oh, I've gotta hit those numbers. Oh, I'm doing terribly.'"

Amazon's brand of modern-day Taylorism is ingenious, when you fully unpack its clockwork perfection. It's a mix of surveillance, measurement, psychological tricks, targets, incentives, sloganeering, Jeff Bezos's trademark hard-charging attitude toward work, and an ever-growing array of clever and often proprietary technologies. Taken as a whole, this system is unique enough in the history of work that it deserves its own name: Bezosism.

After Taylorism, Fordism, the Toyota Production System, and dozens of other sectarian management philosophies, it has come to this.

There have been many attempts to describe Bezosism, or at least its constituent parts. That I find all of them insufficient in their scope and descriptive power is one reason you're reading this book. Those who have come the closest talk about "digital Taylorism," "neo-Taylorism," "management by algorithm," or "algorithmic despotism." But all of these descriptors miss the most critical element of Bezosism, which is that while it comes from a long line of other management systems, and is being developed in parallel with them, it is truly native to Amazon. No other company on Earth today is as well resourced, talent-rich, ruthless, or fast-moving as Amazon when it comes to tightly coupled systems of machines and humans, AI and wetware, robots and bodies.

Taylorism was developed by many, but it is deservedly known by the name of its most vocal proponent. Fordism was the product of a team of clever and hardworking engineers, but it is known by the name of the man without whom it would not have coalesced. The Toyota Production System evolved by trial and error on the part of a handful of the firm's leaders, but it is known by the name of the company that pioneered it. In turn, Jeff Bezos deserves credit for and should be identified with Bezosism.

At this very moment, Bezosism is diffusing through the world of work, rewriting the source code of the global industrial machine. If it proves as popular and durable as the systems of organization on which it builds, it will be, after the company he built, Jeff Bezos's most important legacy.

Depending on how the company practicing Bezosism wields its power, it can be benevolent or sinister or both. In its darker moments, Bezosism is what Paul Adler, a professor of management and organization at the USC Marshall School of Business, calls "despotic Taylorism." Labor practices like Amazon's are not the inevitable outcome of today's systems of production and management, he notes, but what can happen when workers lack the power to push back against management.

But how does Bezosism work? Is it merely more of what came before—a more intense version of mass production, at a faster pace—or different in kind?

The first thing to understand about the highly evolved management system known as Bezosism is that nothing about it is accidental. Its every spandrel was deliberately built and refined by the leaders at Amazon, even aspects of it that many who have worked at the company find alienating. The second thing to understand is that Bezosism doesn't exist without the technology that enables it. And the last key point about Bezosism is that Amazon's leaders do not seem to have fully come to grips with the effects of the system they've created. Maybe nobody has.

Amazon has hit upon, and in some cases borrowed, methods of motivating its employees in which its managers need never raise their voice or dress down a subordinate. In the game that all managers play, of carrot and stick, incentive and rebuke, Amazon's managers have the opportunity to always appear sympathetic, to always be on the side of employees. Inside Amazon's warehouses, should they choose to, all the humans get to be united in the face of the real taskmaster in the building: the algorithm.

It was the algorithm, you might say, that led to Austin's carpal tunnel syndrome.

"There's no way somebody can do the same job over and over again and not have their body crap out on them in some way," says Austin.

Owing to Amazon covering all its full-time employees' health insurance from the day they start the job, Austin was able to see a doctor about his repetitive stress injury before he left. She told him how to take care of it and gave him splints for his wrists. What ultimately resolved the issue wasn't anything the medical system could offer him, however—it was that he quit. While Austin had always planned to work at Amazon for only a short time, his injury made staying on impossible. In a way, he was one of the lucky ones.

Every time Emily Guendelsberger picks up a suitcase with her right hand, pain shoots through her elbow. It's a souvenir of her time in an Amazon fulfillment center, where she worked while researching her book *On the Clock: What Low-Wage Work Did to Me and How It Drives America Insane.*

It's apparent that Emily's book drives Amazon's press people up the wall. Their official response, when asked about it during her post-publication media tour, was to point out that while she claimed to have worked at an Amazon fulfillment center for nearly a month, she actually only worked a total of eleven days. This is true—she got the flu during her month at Amazon and had to take time off—but also an admission that a reasonably healthy woman in her midthirties can acquire a permanent, if not exactly debilitating, injury by working as an Amazon associate for as little as three weeks. (Workweeks at Amazon typically last four days.)

Then there was the matter of the work itself. In its videos and tours, Amazon touts its automated warehouses, the ones with the pick towers and robot floors full of Mick Mountz's mobile shelves atop Kiva drive units. *Look*, they say, in so many words, *these robots are making everyone's jobs easier.*

"The robots change the work so they allow us—people don't have to walk as far, which is a complaint that we've heard in the past," Jeff Wilke, whose official title at the time was CEO of worldwide consumer at Amazon, told a film crew from the PBS documentary series *Frontline* in 2019. "They make the job safer," he added. Wilke is the person most responsible for the design and function of Amazon's fulfillment centers. Until he announced in August 2020 his plan to retire early the next year, he was also heir apparent to Jeff Bezos.

What Amazon rarely mentions, and only ever in passing, is that the overwhelming majority of its warehouses are nothing like the newest, shiniest, most advanced of its facilities, the ones where the company admits film crews and gives tours. According to Amazon, as of early 2019, when the company had 175 fulfillment centers worldwide, only twenty-six of them included robots. That's just 15 percent.

And while they are by far the largest facilities Amazon operates, fulfillment centers are only a fraction of the total population of the company's warehouses. It's hard to get a definitive census of these facilities, but estimates from logistics consulting firm MWPVL International peg Amazon's global footprint at 1,511 logistics hubs of one sort or another as of December 2020. That includes Prime Now hubs, which fulfill same-day orders; outbound sortation centers, which are critical to directing packages from fulfillment centers to the smaller warehouses where they're loaded onto Amazon-branded delivery trucks; inbound sortation centers for directing goods into fulfillment centers; delivery stations, which are the final stop on a package's journey before it's placed on a delivery van; airport hubs; fresh-food fulfillment centers; and Whole Foods retail distribution centers.

While some of these facilities do have robots—Amazon's newest robots power some of its most recently constructed sortation centers—the rest lack Kiva-style robots. That means Amazon is running more than a thousand facilities that are little different from traditional warehouses. At these facilities, the warehouse management software may be sophisticated, but the actual work is still done by tools that have changed little in the past fifty years—forklifts, pallet jacks, and hundreds of thousands of humans walking, lifting, sorting, unpacking, packing, and loading.

It was at one of these less-automated fulfillment centers, just outside Louisville, Kentucky, in which Emily found herself working. In these fulfillment centers, all goods are stored on a vast array of stationary shelving laid out in a grid, like the streets of Manhattan. Pickers must walk to the location of every item, directed by a portable bar code–scanning gun with a display on its top.

"Amazon was safety-focused to an almost annoying degree," says Emily. Her training included how to lift things properly, the importance of staying hydrated, and why you never stick your hands in a conveyor. Former Amazon workers have recounted the tedium of hours of training in which they were forced to listen as instructors read aloud lists of safety rules, maxims, and exhortations.

According to Emily and others I spoke to, much of that safety training goes out the window when you've got to "make rate," which means matching or exceeding an hourly count of how many objects you're supposed to pick, stow, or pack. One day, in a state of utter exhaustion, Emily kneeled to grab something off a lower shelf and discovered to her horror that her legs had simply stopped working. Anyone who has ever pushed themselves past the point of complete exhaustion knows the sense of surprise and alienation that can overwhelm you in a moment like this.

There Emily sat, sobbing. No one saw her and no one came to help. Amazon warehouses that lack robots are so vast that associates rarely bump into one another.

Amazon's logistics hubs are incredibly diverse, their inner workings dependent on their functions and vintage. The pace at which its associates must work varies by facility and time of year, especially around the holidays, when the company may go so far as to rent out additional space from other warehouse providers to handle the increase in demand. Whatever the facility, Amazon argues that the pace at which its associates must work is fair. At its fulfillment centers, the primary way the company dictates the speed of work seems, at least on paper, to be quite reasonable.

On a nonpeak day, the rate a picker or stower must hit in these fulfillment centers is the twenty-fifth percentile. That is, they must be faster than the bottom 25 percent of their coworkers to avoid the threat of being "written up," which can eventually lead to being fired.

Here's where Amazon's comprehensive worker surveillance systems enable the company to monitor its employees from one moment to the next, giving them the proverbial spur when they slow down or allowing management to reward them if they exceed expectations. In a warehouse like the one in which Emily worked, every time she picked an item, the countdown clock on her scanning gun would reset, telling her exactly how long she had to walk to the next item. That time was determined by an algorithm that, in theory at least, knew how long it should take her to walk to the shelf holding that item.

In the minority of Amazon's fulfillment centers that have robots, like the ones in Shakopee, Minnesota, or Baltimore, Maryland, there is no clock, just an overall pick or stow rate calculated for each worker at their robot-fed pick and stow stations.

Imagine the delight of Frederick Taylor or Henry Ford if they could know, to the millisecond, how long it took every worker to complete a task, every day, in every facility they owned. Imagine what time and motion experts Frank and Lillian Gilbreth could have accomplished had they been able to discard their film cameras and replace them with millions of hours of video captured from the digital cameras that watch every station at many of Amazon's fulfillment centers. Imagine how much additional just-in-time efficiency in inventory levels, capital allocation, and automated reordering Taiichi Ohno and Eiji Toyoda, creators of the Toyota Production System, would be able to extract from a system that knew the precise moment an associate plucked an item from a shelf and sent it on its way.

Amazon borrows directly from all of these systems, but the company's special twist, that floating rate that all workers must hit, is what makes its facilities particularly Darwinian.

Management theorists have known for some time that forcing bosses to grade their employees on a curve is a recipe for low morale and unnecessarily high turnover. The white-collar version of this management practice, known as "stack ranking," was pioneered at General Electric in the 1980s. Later it came to be known as "rank and yank" because it forced managers to give negative performance reviews to some employees in their teams, even if all of them were generally excellent and the differences between team members were slight.

The original idea, favored by Jack Welch, former CEO of GE, was that every company should aim for a certain level of turnover, whatever the consequences. The system was rife with perverse incentives. Peers who sabotaged others' work could save their own jobs; managers might hire less-capable people on their teams to keep from having to fire existing employees whom they favored. Despite the system's drawbacks, Welch's influence was so far-reaching that stack ranking was adopted at many of today's tech giants, where it wreaked havoc on morale and productivity for decades. Eventually, its negative effects became well known enough to make the practice a liability at companies chasing workers whose specialized talents made them scarce, such as engineers. In the mid-2010s, companies including Google, Microsoft, and Amazon abandoned it.

Arguably, Amazon's system of making rate in its warehouses is the blue-collar equivalent of stack ranking, and it has similar mechanics to the rank-and-yank systems so despised by white-collar workers. According to many current and former employees, it also has similar effects on morale at Amazon. That leads to people quitting jobs in fulfillment centers at very high rates in their initial weeks on the job.

A critical difference between the stack ranking of white-collar workers and the rate system imposed on blue-collar workers is that while companies can ill-afford to lose employees whose skills are in demand, shedding workers who are easily replaced is often a company's explicit goal. This is almost certainly why one of these systems has died out and the other persists. This practice is hardly limited to Amazon. It's common across a huge swath of jobs requiring little training, as Emily discovered when she also worked at a call center and a McDonald's.

According to the accounts of many workers who spent years at Amazon, the rate they must hit from one day to the next fluctuates, but overall it's gradually going up. This results in all kinds of perverse incentives.

"The percentile curve is just based off of what everyone is doing right now," says Tyler Hamilton, the picker and stower at Amazon's Shakopee, Minnesota, fulfillment center. He's describing the curve that results when managers plot the rate at which every worker in a fulfillment center is working. "It's not set off of, 'OK, here's a group of people doing this according to exactly how they're supposed to do it to ensure quality and safety.'"

Frederick Taylor and his disciples could only measure the performance of workers over a limited span of time and then create an idealized model of the methods and speed with which they should be doing their job. But Amazon is able to make constant measurements of how every one of its workers is actually doing. In theory, this could lead to a more generous system of management, one in which the pace of work reflects the overall health and motivation of all workers in a warehouse.

In practice, a system in which most associates are in the middle of the bell curve of the distribution of performance is a system in which most workers in an Amazon warehouse are constantly in danger of losing their jobs, and they know it. Their only incentive is to continue to go faster, or else risk losing their benefits, paycheck, and any seniority they've accrued.

To understand why this is the case, consider first that people have bad weeks. We all get sick, but for its hourly warehouse workers, Amazon's sick leave is unpaid. Even in normal times, many employees can't afford the lost wages that come from not showing up at work. All unpaid leave for any purpose is also limited to ten days a year, and anyone who is responsible for a parent, friend, neighbor, spouse, child, or even just a geriatric dog must conserve that time for real emergencies. Taking more than your allotted time off is, of course, also a way to get fired.

Now think about the shape of a bell curve describing the statistically "normal" distribution of workers at an Amazon fulfillment center. Because of the relatively steep slope of the left side of that curve, pretty much everyone in the middle of the distribution—the majority of workers at a fulfillment center—is always not that far above the rate they must make to avoid losing their job. Many of them are, in other words, in a constant state of worry about whether or not they can endure the job they're doing.

Thus, a system that seems fair instills a level of fear in workers that forces many of them to push themselves to the very limit of their physical and mental stamina, and beyond.

"So if there are people who cut corners, if there are people who take tons of coffee and tons of energy drinks to go faster, that raises the cumulative rate," says Tyler. "Meaning, if you want to keep up with the average, then you have to cut corners and drink coffee and energy drinks at every break."

Cutting corners and getting juiced on caffeine isn't just something people do when it's Prime Day or peak season. For many, it's what they do all the time. "I mean, the coffee is free out of the machines," adds Tyler. Another thing that is free at Amazon warehouses is ibuprofen, available from no-cost vending machines scattered throughout the warehouse.

Because bathroom breaks are the fastest way to mess up your rate, you can also get Imodium from AmCare, the on-site medical facility. In moderation, all these things are fine, but if they're how work gets done every day, they're not sustainable. As any sleep expert on the planet will tell you, with caffeine there is no free lunch.

When modifying the physiology of your body isn't enough, there's always the option to chuck your safety training. Every pick and stow station has a step stool so that workers don't have to stretch to reach high shelves, but to make rate when they're in hour six of a ten-hour shift and starting to flag, they sometimes ignore them. Then there's the classic mistake of not using your legs when reaching lower shelves, something an associate must do hundreds of times in a shift. Just bending over can be more expedient, but it risks other sorts of injuries.

Then there are the ways employees have figured out for boosting their rate by sacrificing quality. "Machine-gunning" is when they shove as many small objects as they can into a succession of bins, without bothering to check that all of them are intact or match the description listed in Amazon's databases, says Tyler.

A central tenet of the management practice of *kaizen*, a part of the Toyota Production System in evidence everywhere at Amazon (and which will be discussed further in chapter 17) is that mistakes must not be passed down the line. To this day, many associates in Amazon fulfillment centers have their own version of the famous kaizen "andon cord" they can pull when there's an issue at their station. In a manufacturing facility, this is supposed to stop the entire production line, but in an Amazon facility, it just alerts other workers in problem-solving roles that their help is needed.

With the all-important rate looming over them, some workers pass errors down the line, says Tyler. Those errors might be mislabeled or damaged products, or items stuffed in the wrong bin. Amazon itself might not be fully cognizant of these issues, but they are all too familiar to automotive plant managers in the United States. Beginning in the 1980s, these managers were forced by competition from Japanese automakers to transition from Fordism, where mistakes are passed down the line to a "rework" area, to the kaizen system, in which mistakes aren't supposed to happen at all.

Without its leaders even realizing it, Amazon's demand for a furious working pace could actually be lowering productivity within its facilities. It's what happens when the only incentive is to meet certain metrics. Employees quickly figure out what they can sacrifice to meet those numbers, leading to externalities that lie outside what management measures and prioritizes. Items missing from bins, damaged goods that lead to refunds and lower customer satisfaction, the loss of employees who might otherwise stay with the company and climb its ranks—all are issues Amazon could be having without being fully cognizant of the trade-offs its drive for productivity has led to.

It's difficult to quantify the impact of Bezosism on workers, but some have tried. The company refuses to release data on injury rates at its fulfillment centers, and it is not obligated to under federal law. The exception to that rule is that current and former workers at a facility are entitled to that data. Enterprising journalists at *The Atlantic* and the Center for Investigative Reporting (CIR) organized a network of such workers, gathering enough information to compile a database of the injury rates at twenty-eight Amazon warehouses.

At the worst of these places, injury rates were four times the industry standard for warehouses. Amazon has argued that its injury rates only look high because the company's safety culture means that it obsessively documents incidents in a way that its competitors do not. Workers within its facilities have countered that its singular focus on reducing accidents leading to acute injury is in some cases having the opposite of its intended effect. If someone is injured, some managers will encourage them not to report it, for fear that it will be a black mark on their own records.

The facility where Austin worked and in which he had an experience so positive that, despite his exhaustion and repetitive stress injury, he wrote a thank-you note to his bosses after he left, is the one with the absolute lowest injury rate of all the Amazon facilities surveyed.

At the other end of the spectrum are facilities like the one in Eastvale, California, where 422 injuries were recorded in 2018 alone. Incidents in which workers come away from Amazon having suffered permanent disability or worse are what make headlines. But these are not the incidents that worry observers like Emily the most. Rather, as she experienced firsthand, and as countless Amazon workers the world over have related in interviews and through their protests of working conditions, the most pervasive issues that come from working at an Amazon warehouse are the ones related to being treated like a robot.

"There's a difference between hard work and chronically stressful work," says Emily. The anxiety inherent in having to make rate, like the ways that truck drivers must push themselves in order to eke out a living, is a theme that comes up again and again in interviews with current and former Amazon associates.

At the 2019 *Wall Street Journal* Tech Live conference in Laguna Beach, California, *Journal* editor-in-chief Matt Murray asked Jeff Wilke about news reports of injuries and other issues in Amazon's warehouses. Wilke replied breezily that with a company that employs as many people as Amazon does, it's always possible to find enough anecdotes to fill an article.

I remember the moment distinctly, because Wilke's answer struck me as a rare combination of both plausible and glib. Now that I've read dozens of interviews with former employees, gone down the rabbit hole of "Why I left Amazon" testimonials on YouTube—where such accounts are practically their own genre—and interviewed my share of current and former employees, all I can think is that when most of the people who have worked for your company in a blue-collar role testify that it was, at best, an exceptionally difficult place to work, it's much harder to dismiss their claims.

▶ ▶ ▶

Amazon's internal culture, from Jeff Bezos on down, is by most accounts fundamentally Darwinian. For the entry-level associate, one expression of this philosophy is that the company has all but abandoned interviewing new hires. Instead, once applicants for jobs in Amazon's warehouses pass a drug test, they're given a date to show up for orientation. There, they are told about the physical requirements of their jobs. The work may require walking more than ten miles a day, they're told, and also lifting heavy boxes over and over again, while working to a high standard under tight deadlines. Applicants are encouraged to leave now if they think they can't cut it.

This turns out to be a not-very-effective mechanism for winnowing Amazon's applicant pool. Everyone I talked to about their first months at Amazon said that the attrition rate they witnessed was greater than 50 percent in the first two months.

Austin recalls that when he joined Amazon, there were eight people in his group of trainees. After the first night, two people didn't come back. The second night, two more left. One morning a few weeks later, another one of his coworkers told him, "I'm not coming back. I can make fifteen dollars an hour without killing myself."

Much has been written about the high cost of turnover at businesses, but this seems only to apply to ones that invest in the training, onboarding, and development of their employees. For the sort of firms that hire workers regardless of their level of education, whom they train in a matter of hours, turnover is potentially an advantage.

This is hardly a realization original to Amazon. As I mentioned in the chapter on trucking, even an industry that requires its workers to spend a month or more training can sustain an annual turnover rate close to 100 percent. Retail as a whole has a turnover rate of around 60 percent, and discounters like Walmart, known for squeezing their employees no less than they squeeze their suppliers, exceed that at some stores. The fast-food industry in general has a turnover rate greater than 100 percent.

It's not clear what the turnover rate is at Amazon, and according to accounts of former employees, it varies a great deal depending on how long someone has been with the company. Amazon claims that overall turnover among its full-time workers dropped in both 2018 and 2019. Complicating the matter is that Amazon also has an annual influx of seasonal workers who come and go quickly, including 200,000 additional associates worldwide in 2019 alone.

When the cost of recruiting, training, and integrating new workers is low, high turnover has a number of advantages. The first is that the company can provide only a minimum of pathways into management. This saves firms money on payroll by maintaining a high ratio of front-line workers to managers. This is especially true in firms that manage by algorithm. One of the ironies of Bezosism is that it's not the jobs of blue-collar workers that are disappearing, but the jobs of middle managers.

Another way turnover can help companies like Amazon is that it makes it almost impossible for workers to unionize. Amazon, like Walmart, is notorious for discouraging workers from forming unions, but to date the company's most potent weapon in the fight against employees gaining any leverage over management is that so few are sticking around long enough to organize.

▶ ▶ ▶

It's generally an article of faith among even the most dispassionate observers of Amazon and its operating system, Bezosism, that all this speeding up of work is the inevitable consequence of more and more technology. But, to borrow a phrase analyst Ben Thompson originally used to describe the internet, technology is an amoral force that reduces friction, not an inevitable force for good or ill.

When Kiva's engineers and managers first started rolling out their systems in warehouses belonging to companies like Walgreens, employees loved them, says company founder Mick Mountz. And why wouldn't they? Employees went from walking ten or more miles a day to retrieve items for delivery to walking almost none, because the inventory came to them, atop robots.

But imagining that a new technology that can make someone more productive will ultimately mean they have to do less work is a classic mistake. The reality, demonstrated again and again throughout history, is that new technologies mean humans end up doing nearly the same number of hours of work but in a different way. Famously, John Maynard Keynes proposed in a 1931 essay, "Economic Possibilities for Our Grandchildren," that thanks to automation, by 2028 everyone would be working about three hours a day.

Clearly, that didn't happen. If you exclude vacations but assume a person gets weekends off, Americans actually average about six and a half hours a day of paid work. Now add in the additional four and a half hours of unpaid domestic labor women do every day, or the two hours a day of domestic labor done by the shiftless men of the world's rich countries.

What Keynes got wrong is what lots of people get wrong about the relationship between new technologies, increased productivity, and jobs. New technologies don't eliminate jobs. As living standards rise, humans find new ways to consume more. Just as important, every time we automate a task, we tend to use more of that product or service, in combination with others, to accomplish some other more complicated or difficult end. Consider, for example, how telephone switchboard operators were replaced with circuits, without which we could not have the internet.

Every time there's a recession—or a global pandemic—and people are forced to change their ways, or switch to a cheaper and better alternative, the process of disruption goes into overdrive. The result is that the demand for new goods and services goes up at the same time that humans suitable for providing them are looking for jobs. And some innovative new company, such as Amazon, figures out how to meet that shift in demand by scooping up all those underutilized workers.

As Amazon itself puts it, "The fulfillment centers that have robots often have higher employment numbers because inventory is moved at a faster pace, which requires extra associates." And this is hardly the first time in history that a new technology required vast numbers of new unskilled workers. For example, historians broadly agree that before Eli Whitney's invention of the cotton gin, slavery was on its way out in the United States. But the combined appetites of an automated gin and England's rapidly industrializing spinning and weaving industries created a huge demand for enslaved persons and made their continued abuse and bondage an enormously profitable enterprise for white Southerners.

A worker using the Kiva system in its early incarnations would typically triple their output, say from an average of 100 picks an hour to 300, says Mick. But it wasn't as if the Kiva-using companies then reduced all their warehouse employees' hours to a third of what they once were while paying them the same wage. Instead, Staples and Walgreens, both early customers of Kiva, leveraged their workers' increased productivity to increase the output capacity of their warehouses, store and ship a wider range of products, shorten the amount of time required to fulfill an order, and ultimately either lower the cost of their services, increase their profits, or both, as capitalists have done since the dawn of the corporation.

Frederick Taylor made the same error when he wrote that his system "would insure higher wages and make shorter working hours and better working and home conditions possible." Louis Brandeis, best known as a champion of labor, repeated the error when he advocated for Taylorism. People who are invested in dreams of technology easing our burdens by giving us more power over the world often forget that technology in no way changes the power structures that govern it.

At Amazon, the "rate" is the purest expression of both the company's goals and who is in charge of defining them. Amazon's leaders and spokespeople like to talk about how automation makes the job of an associate easier. But they seem unable or unwilling to imagine that the increased demands of that automation on the associates could be grinding them down both physically and psychologically. Nor do they seem willing to entertain the notion that using ever-finer slices of workers' cognitive and physical capacities, robbing them of the chance to use their other talents, creates problems of its own.

While the data is hardly definitive, it's notable that in the *Atlantic/ CIR* investigation, it appears that injury rates are actually *higher* in facilities that include robots. This trend was later confirmed by a trove of additional injury reports and internal documents obtained by Will Evans of the CIR, published in October 2020.

In one case, at a warehouse in Tracy, California, the serious-injury rate went from 2.9 per 100 workers in 2015 to 11.3 in 2018, after the introduction of robots. This makes sense when you consider the nature of the work—faster, more repetitive, and spurred by an algorithm that keeps the pick, stow, and pack rates floating near the maximum that can be accomplished by people toiling in Amazon's warehouses. As one worker at the facility told Evans, "Before robots, it was still tough, but it was manageable. [After,] we were in a fight that we just can't win."

I asked Mick to comment on the injury rate at facilities with Kiva robots, and he said that in the original design of the Kiva system, "we always pointed out the human is in control of the machine, not the other way around. We'd say, this is not the Lucille Ball episode where she's on the chocolate line. That's the old-school way, where the automation is driving the human. In the Kiva system, the human drives the automation. If it takes me twenty-seven seconds to pack out a wedding dress, that's how long it takes. It's only when I scan an item that a pod leaves and another shows up. Whatever pace you're working at is the one that it feeds you at."

One element of the Kiva system that Amazon has retained to this day is that Amazon's warehouse management software knows exactly how quickly every worker is stowing or picking. Those numbers are crunched in real time by a system developed by Amazon, called, appropriately, ADAPT. It directs more robot shelves to faster workers, maximizing the flow of goods through the warehouse. It also rewards the most efficient humans with the ultimate prize—more work.

The rest of Mick's thinking about how to build a humane system of collaboration between humans and robots appears to have gone out the window, if it was ever truly part of how Amazon and other companies used these systems. It turns out that how fast everyone in the warehouse must go has nothing to do with the technology itself. It's a management decision.

"Whether a customer, be that Amazon or Walgreens, says you have to pick 800 items an hour or 300 an hour, that's a function of the type of inventory you're handling, and management philosophy," says Mick.

If one of the hallmarks of Fordism and mass production is the process of making tasks so simple that anyone can do them with almost no training, a process known as "de-skilling," then the factory-like supply chain of Amazon is Fordism on a whole new level. Indeed, without de-skilling, Amazon could not exist.

De-skilling is what makes it possible for Amazon to continue to function despite relatively high rates of turnover. It's also what makes it possible for the company to hire hundreds of thousands of seasonal employees every year and, whatever their background, turn them into productive associates within a day or two.

The spinning jenny, Jacquard loom, and numerical machine tool, all milestones in the industrialization of manufacturing, took knowledge that used to be in the heads of skilled craftspeople and embodied it in a machine that made them redundant. Today, automation does this and more: it makes possible things that no human could accomplish without it.

Amazon's warehouses embody this to an extreme degree. Only a hive mind as vast as the millions of lines of code and massive neural networks possessed by Amazon could figure out how to get goods out of any given fulfillment center and onto the doorstep of a customer within a day. The closest predigital humans ever came to this feat was the mobilization of factories and material carried out by the United States during World War II, and that was accomplished by millions of human minds moving mountains of paperwork over the course of years.

The tools, dies, and jigs that made Fordism and mass production possible presaged more exotic forms of automation. Today, we have predictive analytics that can tell companies how much of a thing to produce and ship across the ocean, demand-prediction and load-balancing algorithms to determine how much of that inventory should go to each fulfillment center, AI-controlled movable shelving, the Amazon Fulfillment Engine, dynamic route planning for delivery vehicles, and on and on down through the supply chain. No human participating in any part of this leviathan could ever have a full understanding of the decisions being made by machines that make this system possible. The front-line associates at Amazon are both beneath the algorithm and managed by it. Every routine task they undertake is decided not by them, not by any human, but by a nonorganic intelligence.

By its very nature, de-skilling leads to workers doing the same thing over and over again. In manufacturing, the rate at which a worker has to repeat a process is known as the "cycle time." In the early days of car manufacturing, when each team of workers responsible for final assembly would build an entire car from start to finish, their cycle time could be measured in weeks or even months. Ford's assembly line reduced that cycle time to minutes; in an Amazon warehouse, the cycle time is seconds.

While being human has always involved some amount of tedium, the duration and intensity of tedium that must be endured by an

Amazon associate is something rarely seen outside of mechanized slaughterhouses and certain kinds of factories.

Some associates I interviewed were fine with this kind of work. Robert Taylor had worked as an associate in Amazon's Baltimore fulfillment center for three years when I interviewed him. His boss, Tyler Walter, described him as a "very fast picker." I asked Robert what his goals are at Amazon. "I just come in every day and work harder," he said. "Amazon turned me into one of their robots," he added, laughing.

Plenty of associates are not so happy being robots. One frequent complaint is that, in addition to the repetitive and mind-numbing nature of the work, they have very little opportunity to interact with anyone else, and aren't even allowed to listen to music. Amazon has said for years that its no-headphones rule is a safety issue and that it's working on the problem.

It's not as if the challenges of repetitive factory-type work are unknown. Workers who had to endure the accelerated pace and repetition of Ford's assembly line described their discombobulation at the end of a workday as "Forditis."

Ford solved the unsustainable levels of turnover at his factories that resulted from Forditis by more than doubling his workers' wages to the then-unheard-of $5 a day. Amazon, on the other hand, painted its $15-an-hour wage as an act of pure noblesse oblige. In an October 2018 letter, Jeff Bezos said Amazon had "listened to its critics" and "decided we want to lead." But the company has also noted that the shift to $15 an hour hasn't exactly hurt recruitment. Amazon has said that in the month after it announced its $15-an-hour wage, it received approximately 850,000 applications for seasonal positions. The company also noted that this was twice the number of applications it received for such positions in the month before that announcement. Which means that at least in the run-up to the holidays, Amazon clocks nearly half a million applications a month from workers seeking temporary positions. This doesn't even include applications for permanent positions, which the company doesn't disclose.

These numbers get to the heart of how Amazon is able to demand so much from its workers. By being a gigantic, highly visible company that offers above-market wages, plus health-care benefits starting on day one, Amazon is able to attract huge numbers of applicants. And because it relentlessly pursues automation to make the jobs of its workers ever simpler and more error-proof, it can attract and productively employ huge numbers of potential associates regardless of their level of skill or education.

The genius of ride-sharing companies Uber and Lyft (not to mention delivery start-ups Postmates and Instacart) is that thanks to two-sided marketplaces run by software, route-planning algorithms, and GPS, all you need to know to work for these companies is how to drive a car. In the same way, the genius of Amazon is that all you need to possess in order to work in one of its warehouses are sensorimotor skills most people acquired by the time they're in grade school.

De-skilling and higher wages aren't just a good way for Amazon to staff up as it rapidly grows. They're also, arguably, absolute necessities in a tight labor market. Just before Amazon announced its $15-an-hour wage, the official unemployment rate in the United States dipped below 4 percent. Once the pandemic hit and Amazon needed to hire at a record pace, de-skilling was even more important, because it meant the company could take on almost anyone.

Even with jobs that require little in terms of skills, Amazon faces the challenge that as it expands its network of fulfillment centers, many others are doing the same, and often in the same areas.

Leif Jentoft is cofounder of RightHand Robotics, a company aiming to replace as many of the associates in an Amazon warehouse as possible with automation. Being in the business of supplying picks and shovels to the rapidly expanding e-commerce industry means he is uniquely privileged to see the inside of a great many distribution and fulfillment centers. At some point, he noticed how many of them were clustered in certain areas. "E-commerce alley" is what he calls the chain of warehouses strung out along interstate highways on the East Coast.

"From Kentucky up through Pennsylvania, if you create a map of a transportation network where you can reach people with two-day delivery, there's only a few areas where you can do that," says Leif. "All the different retailers are running the same transportation optimization software, which means you end up with these distribution centers all in the same areas."

That can lead to labor wars for warehouse workers, at least in some places. Even though these jobs are unskilled, because warehouses are such big employers, and the work can be intensely physically and psychologically demanding, there are only so many people in commuting distance of these facilities that these firms can draw from.

In order to draw on a big enough pool of workers, Amazon frequently places its fulfillment centers in areas that have high levels of unemployment. Many are in places where America once made things, including the Rust Belt and California's Inland Empire.

This is the central irony of the modern supply chain and Bezosism in general: in parts of America where people once made things in factories, their children now work in factory-like e-commerce fulfillment centers. Those workers are still touching many of the same goods, but now they're manufactured overseas. What's more, these fulfillment centers employ more sophisticated and technologically advanced versions of the same labor and management practices as the factories they replaced. In America and in rich countries the world over, for many workers, the warehouse is the new factory.

When you look at these jobs through this lens, $15 an hour doesn't seem like nearly as much. It might be a high wage when compared to working an entry-level job in the service industry, but if you compare it to a unionized, $40-an-hour factory job that was outsourced in the name of economic advantage, it's clear that for many Americans, manufacturing dollars have turned into supply chain cents.

These fulfillment center jobs are, for now at least, impossible to outsource like the factory jobs that came before them, because next-day delivery depends on the proximity of warehouses to the people who consume their contents. That doesn't mean these jobs are safe forever, though. Roboticist Rodney Brooks once described his motivation for creating robots that could work alongside humans like this: If the trend of rising wages in developing countries continues, manufacturers will eventually run out of pools of cheap labor. At that point, if a company wants to remain competitive, the only way forward is more automation and more robots.

This doesn't mean humans are in any reasonable span of time going to be "replaced" by robots in the labor market as a whole. What all the robots, automation, and AI that are a part of Bezosism have accomplished isn't, manifestly, the disemployment of millions of people. There is no asterisk attached to Amazon's claims that even as it has added more than 200,000 robots to its ranks, it has also added more than 300,000 people. Talk of the "retail apocalypse" and the collapse of many old standby brick-and-mortar brands notwithstanding, what's happening to work isn't that it's going away.

Rather, as Emily Guendelsberger, the journalist who toiled in an Amazon warehouse, puts it, what we're seeing is the rise of "cyborg jobs."

"These are jobs where you're held accountable to robotic standards of efficiency and productivity and you're expected to quash all your human failings," she says. "Being a human is regarded as a failing when you're in competition with machines and algorithms for your actual job."

Those human failings include having to go to the bathroom, having to take care of your kids outside of work, and "not being able to just do the same exact thing for eleven hours without talking to anybody, without music or any sort of distraction to keep your mind off the fact that this is an incredibly boring and monotonous and repetitive job," she continues. "It's putting a large portion of the American workforce in direct competition with these machines."

Again, this mode of work is hardly new. Ford himself recognized and addressed it in his autobiography. And in no small irony, the mechanization of production of Ford's era was the inspiration for the 1920 work by Czech playwright Karel Čapek that gave us the word "robot." The android-like "robots" of that play were intended as slaves for their human masters, but they eventually rebel against their creators. Čapek's original definition of a robot—a thinking person squeezed into the mold of an automaton—gives extra resonance to the protests of Amazon associates in Europe, Brazil, and the United States who have held up signs saying WE ARE NOT ROBOTS!

In Ford's era, these kinds of jobs were confined to the factory floor. But now, notes Emily, "the assembly line is everywhere. You can no longer vote with your feet because it's going to be just as bad at Amazon as it is at Walmart, or at McDonald's."

There are some escapes, of course. Upward mobility may be at historic lows in the United States, but education and hard work still propel some out of these jobs. Amazon itself announced in July 2019 that it would spend more than $700 million to "upskill" 100,000 of its employees, propelling them into roles including "data mapping specialist, data scientist, solutions architect and business analyst, as well as logistics coordinator, process improvement manager and transportation specialist."

I asked associates about this program and the opportunities for training at Amazon in general. Associates in the Baltimore fulfillment center said that the most popular course of study for associates there was a commercial driver's license. This begs the question of whether those acquiring one have any idea how similar being a commercial truck driver can be to being a picker or stower, even if it comes with a different set of stresses.

Associates elsewhere said they had no idea how anyone would have the energy to go to school while clocking forty or more hours a week in a job as grueling as warehouse associate. Many of the skills and jobs Amazon mentioned seem geared toward Amazonians who are already in some kind of white-collar job. For example, Amazon already has pathways for IT support staff to transform themselves into the kind of coders who are in short supply.

Chapter 17

FROM JAPAN WITH LOVE

Origins of the Amazon Way

The Amazon associates who reported being most satisfied with their jobs all talked about having some sense of agency. Bosses who listened to them, opportunities to learn new skills, and a sense of ownership over their jobs were all areas in which they gave high marks to Amazon.

Inevitably, this opportunity to improve both their own jobs and the system in which they work is couched in terms that sound like some kind of internal corporate Newspeak, at least until you talk to the people who instituted it. All of it, it turns out, derives from the concept of "lean production," also known as the Toyota Production System and sometimes *kaizen*, though a management consultant steeped in the differences between these related concepts might scoff at a layperson's attempt to use them interchangeably.

The person most responsible for bringing lean production to Amazon, for introducing Jeff Bezos to it and then implementing it across Amazon's supply chain, is Marc Onetto. Marc rose to the level of vice president at General Electric and then moved on to Solectron, one of the earliest third-party manufacturers of electronics for other firms—sort of the Foxconn of its day. (Foxconn makes the iPhone for Apple.)

In 2006, sensing that he needed someone with deep experience in managing complicated supply chains, Bezos tapped Marc to head up

Amazon's online fulfillment operation, which was growing at an explosive rate. His official title was senior vice president, worldwide operations and customer service. In the entire history of Amazon, there have only been four people in this role, while also being the public face of Amazon's fulfillment and delivery operations. First there was Jimmy Wright, former vice president in charge of distribution at Walmart from 1990 to 1998, who came out of retirement for a fourteen-month stint at Amazon from 1998 to 1999. Then there was Jeff Wilke, who by the end of his time at Amazon in early 2021 was tied with Amazon Web Services chief Andy Jassy as Bezos's most senior and trusted deputy. As of 2021, there's Dave Clark, who is also the public face of Amazon's fulfillment and delivery operations. In between Wilke and Clark, from 2006 until his retirement in 2013, there was Marc.

While Marc was at Amazon, the company grew from $7 billion a year in revenue to $70 billion. The majority of that growth was off the back of the e-commerce infrastructure he led.

The foundations of how Amazon would measure its own output and attempt to improve it had been laid by Jeff Wilke, before Marc arrived. Wilke had been trained in the doctrine of Six Sigma, which was invented at Motorola in 1986. In the mid-1990s, it was adopted by General Electric CEO Jack Welch. Six Sigma relies on a system of "continuous improvement," plus statistical methods for evaluating the outcome of manufacturing processes. The goal of Six Sigma, and the origin of its name, is that 99.99966 percent of the time, whatever you're doing to a part or accomplishing in a system, it should come out perfect.

In some aspects, Six Sigma is Taylorism taken to the nth degree. Like Taylorism, one of its flaws is that it treats the human like just another cog in the machine. While in some manufacturing contexts it apparently yielded the results required, trying to apply it to messier situations and more labor-intensive industries is a bit like Secretary of Defense Robert McNamara's approach to "winning" the Vietnam War. Infamously, McNamara demanded mountains of data on how the war was going, all of which only served to obscure the reality on the ground.

Six Sigma fell out of fashion as the fortunes of GE declined, and especially as companies moved on to the doctrine of "disruption," which favored innovation and creative thinking over the relentless pursuit of efficiency.

Executives are always on the hunt for the next management fad, and it was kaizen's turn. Advocates of kaizen, or lean manufacturing, shared with the disciples of Six Sigma an obsession with eliminating defects in the manufacturing process itself, at the moment a thing was made, rather than trying to repair those flaws at the end of the assembly line, as had been common in the Fordist age of mass production.

Six Sigma "black belts" and lean manufacturing consultants also shared a method for accomplishing this defect-free assembly line known as "continuous improvement." On the foundation of this shared understanding, Marc built trust with Jeff Wilke.

"Wilke did a super job in putting some discipline into the system [at Amazon] by doing a lot of Six Sigma work," says Marc. "When I came, I continued that and supplemented it. We're very much on the same wavelength."

From the earliest, frenetic days of Amazon's distribution operation, when the company shipped books out of its first "warehouse"—really a two-hundred-square-foot-basement room in downtown Seattle—the company's breakneck expansion presented all its chiefs of logistics with a near-impossible task. Jimmy Wright brought the first automation to this system that it had ever seen, designing for Bezos a brand-new warehouse full of what was then the state of the art in automated sorting technology. Books in this warehouse were sped on their way by massive tilt-tray sorters known as "Crisplants," for the company that made them.

The problem with these highly automated systems was that they weren't very good at handling things other than books. Wilke responded by working on systems that were less automated but more flexible. This meant hiring more humans to do the job. Thus was born Amazon's strategy of using humans and robots in nearly equal proportion, each assigned to the tasks they do best.

Marc's first long conversation with Jeff Bezos was his interview for what would become his job at Amazon. That interview proved auspicious. In Marc's telling, Bezos explained how "customer obsession" was the central goal of Amazon, and that a key way to fulfill its promise to customers was to deliver goods on time.

This obsession with on-time delivery was etched into the brains of Amazon's leaders by a string of disasters and near disasters. The first was the holiday season of 1998. The company's CFO, Joy Covey, noticed that the number of orders coming into the website was rapidly outpacing the number of deliveries leaving the company's warehouses. This crisis led to an operation dubbed Save Santa, in which all hands took to the company's nascent distribution centers to make sure packages got out the door fast enough.

"So I explained to Jeff what the Toyota Production System was," says Marc. "By eliminating waste, which is defined as what the customer won't pay for, I explained to him that by bringing lean into operations, we would drive customer-centricity in every work station on the shop floor and every process in the logistics operation, thereby assuring that we would deliver the right items at the promised time, and at a lower cost to the customer."

Here the durability and broad applicability of lean becomes apparent: From its invention at Toyota starting in the 1960s through its adoption by U.S. auto manufacturers in the 1980s, it was a discipline for people who ran factories. But as management consultants latched on to it and engineers who understood it took on leadership roles at other companies, it spread. Every endeavor undertaken by a group of humans, it seemed, could be "lean-ized," just as everything, Taylor's disciples had declared, could be Taylorized. Soon after it came to the West in the 1980s, kaizen popped up everywhere, in the management of seemingly everything, from health care to call centers.

For Amazon, Marc realized, the equivalents of Toyota's shop floor were the fulfillment centers where the company received, stored, packed, and shipped its goods. But Amazon's shop floor also included the trucks that drove those goods to the fulfillment center in the first place, and the last-mile delivery of goods from the warehouse to the customer.

"All of these are shop floors where things happen, and they have a major impact on delivery of goods to the customer," says Marc. "This impacts the customer as much as the goods available to the customer to buy."

In 2005, just before Marc arrived, Amazon had launched Prime, which guaranteed two-day delivery. It had been a controversial move within the company, because many of Bezos's lieutenants thought either it couldn't be done or that it would be so expensive Amazon would never make money on it. As of the end of 2019, there were 150 million Prime members globally.

In order to achieve two-day delivery, Amazon's leaders had to work backward from the moment an item was delivered to the customer. A large chunk of that time had to go to Amazon's delivery partners. UPS, FedEx, and the U.S. Postal Service had long ago set up large sortation centers devoted to particular regions of the country. If a company like Amazon could get semitrucks full of packages directly to the loading docks of these sortation centers by a certain time, generally the evening of the day before a package was to be delivered to a home or business, these carriers could guarantee it would get to a customer the next day.

That left Amazon just one day to get goods out of its warehouses and trucked to the right sortation center.

"We knew, therefore, what was the last time we could put the box in the truck in order to fulfill the promise to the customer," recalls Marc. "And that would drive the whole picking operation and the whole way we would assign orders to pickers in the fulfillment center."

At every level of the organization, Amazon employees had their own metrics to contend with. Individual associates had to stow, pick, pack, or sort so many items an hour. Area managers who were responsible for dozens of individual associates had to meet targets that were an aggregate of the performance of all their subordinates. And the people in charge of whole fulfillment centers or other varieties of warehouses had to meet targets for their entire facility.

Atop all of those metrics was a single metric for which Marc was responsible. His job was to "meet the promise." The promise was, of course, that a customer's package would arrive when Amazon said it would.

"This was the number one thing Amazon was totally focused on. That was my key metric as head of operations, how many promises did we miss. It was one of the things I would report to Jeff Bezos regularly," says Marc.

Missing a promise was a very bad thing. "We would drive everything in order not to miss that promise," he continues. "It was absolutely essential."

To create a system for Prime members that would meet the company's two-day delivery promise as close to 100 percent of the time as possible, Marc borrowed a concept from lean: the importance of consistency.

Speed is nice, but if sometimes you're getting a package out the door of a fulfillment center in two hours and sometimes you're getting it out in eight hours, being fast is essentially useless. In order to make plans, you must know with certainty how every aspect of your system will perform. You have to, in the language of statistics, reduce the variance of the process you're trying to optimize.

The essence of "continuous improvement" in lean production systems is relentlessly pursuing the root cause of a flaw, whether it's in a manufactured part at Toyota or in a delivery system that's too unpredictable in an Amazon warehouse. Lean prioritizes the pursuit of these root causes above speed, above even getting your work done on any particular day. The idea is that if you systematically pursue the sources of chaos in your operations, you can stamp them out one by one. It's a process that never ends, but the result is a system that is, most of the time anyway, predictable and consistent.

"We'd do a lot of work to eliminate the root cause of a problem and then go into the kaizen, continuous improvement work, to eliminate the reason why things would be unpredictable," says Marc.

For people who don't speak Japanese, the language of kaizen and lean production systems can seem like incantations. Instead of waste, practitioners of lean talk about *muda*. Instead of continuous improvement, *kaizen*. When a production line stops automatically because of an irregularity, that's *jidoka*. *Kanban* is the scheduling system unique to lean production. Instead of "management by walking around," practitioners talk about *gemba* walks. Amazon even created its own word to define its lean system: ACES, which stands for Amazon Customer Excellence System.

Underlying it all is the idea of "just-in-time" manufacturing, which means only ordering what you need when you need it, and making just what's required at that moment. Obviously, that doesn't work unless you have an extremely well-run supply chain. The management of those supply chains is one reason lean production translates so readily to the operations of Amazon.

Once you know the lingo, you can't help but notice that this management ideology is slathered all over everything at Amazon. It's in the company's job postings, its internal literature, the white boards on the walls of its facilities. It's peppered throughout the language of its managers and senior executives.

Tyler Walter, the area manager at the fulfillment center in Baltimore, says his team "does kaizens" all the time, and especially any time a new policy or product rolls out inside the warehouse. The idea is that whenever an associate sees something that could be improved, or an opportunity to increase efficiency, they are to bring it up with their managers.

One example Tyler mentions is that products used to fall out of the soft-sided shelves that ride atop the robotic Kiva drive units in the storage area of Amazon's fulfillment centers. What would be a minor issue in a warehouse staffed entirely by humans can seriously gum up the works of a robot floor, as drive units either run over items or are stopped in their tracks by them. After a kaizen to address the problem, one associate suggested putting elastic bands around each level of the shelf to keep goods from falling out.

When we spoke, Tyler was clearly proud of the fact that area managers like him have the ability to suggest changes in how Amazon operates. But I also heard the elastic-band-around-shelves example from Amazon's head of robotics, Brad Porter. It seems like a founding myth of Amazon's workplace culture as much as a real example of how much power midlevel managers at Amazon actually have. The truth is that however much opportunity associates have to kaizen their workplaces, most of the ways that the machine that is Amazon functions are invisible and beyond the reach of frontline employees. The algorithm that sets the pace of their labors does not take suggestions.

This kind of language also shows up in Amazon's external communications. A job posting for a health and safety manager at Amazon UK includes, under preferred qualifications: "You possess experience of Lean, 5S and Kaizen methodologies." (5S is yet another name for lean.) Another posting for a site manager at Amazon Air in the United States begins: "By leveraging lean principles and kaizens, you will lead continuous improvement initiatives."

In a "fireside chat" in 2012, Jeff Bezos related this story, which by his account occurred in 2007 or 2008, soon after Marc arrived at Amazon:

"I was in a fulfillment center . . . and we were doing a kaizen training session with a Japanese consultant . . . He was from the Toyota school of kaizen . . . He was teaching us some of these Toyota techniques, and he was very fiery. He was very emotive. He would shout at us in Japanese. On about day four, he saw me a little way away, sweeping up some dust in the fulfillment center. He came up to me and said, 'Mr. Bezos! I am all in favor of a clean fulfillment center, but tell me: Why do you sweep? Why do you not instead eliminate the source of dirt?' And I realized there's still a lot of wax on, wax off before we get all the way there."

Amazon retail and logistics CEO Jeff Wilke has talked about the introduction of kaizen to Amazon at the time Marc arrived. Mick Mountz says that his own exposure to kaizen at Motorola informed the design of the Kiva robot system and made it uniquely compatible with Amazon's existing procedures.

"That was one of the things, incidentally, that [Jeff] Wilke loved about Kiva," says Mick. "The Kiva system was built with all lean principles in mind. The way it works is that you only pull the inventory when you need it, and you sort it on the fly, so it had all the tenets of kaizen and lean built into it."

In books and white papers on the concept of kaizen, it's often portrayed as a way to empower workers. Paul Adler, a professor at the USC Marshall School of Business, studied its application at a U.S. auto manufacturing plant in Fremont, California, that operated between 1984 and 2010. Here, he found workers participating in what seems as near to a humane environment as a mass production facility can become. It was, by his account, a "democratic Taylorism" in which workers felt empowered to come up with new ways to be more productive, without doing anything they felt was unsafe, unsustainable, or excessively unpleasant. Accounts of how kaizen was practiced at Toyota itself depict a workplace in which unionized employees with significant leverage were made a part of the problem-solving community that governed Toyota's entire supply chain. At least in its heyday, it allowed these workers to combine their blue-collar work with management-style ownership and thinking about how to improve the production of vehicles.

But for lean production systems to make work better and not just more efficient, it's apparent that workers must either participate directly or feel that their managers are fairly representing them. When I asked entry-level associates if they had heard terms like "kaizen" or "gemba"—terms their managers were quite familiar with—almost to a person they had no idea what I was talking about. Amazon quizzes its employees weekly about whether they feel their work environment is safe, but it seems as though efforts to make them feel empowered, beyond encouraging them to "work hard, have fun, and make history," a slogan common at Amazon facilities, are absent.

This doesn't mean talk of continuous improvement at Amazon is a lie. It's very clear that the company uses this principle, and many others from lean, when its leaders are attempting to improve processes. In our conversation, Marc gave this concrete example of a process improvement he implemented at Amazon, and how he arrived at it:

In his early days at Amazon, Marc saw huge variance in how long it took for associates to get items from the loading dock onto shelves. Sometimes it took as little as twenty minutes, other times it took up to an hour. This was not an acceptable result when trying to stock shelves ahead of the holidays or Prime Day. Looking at the data, the kaizen team noticed that some stowers always got things onto shelves very quickly. "So the kaizen team went to see them and said, 'What do you know that we don't know?'" he recalls.

Goods arriving at Amazon fulfillment centers would come off trucks and be transferred by associates to a desk where they were unloaded from their packaging and marked as received. Then associates put them on carts, from which they would ultimately be stowed. Stowers who knew their business told Marc their secret was that they always selected "good" carts.

"We realized not all the carts were the same," Marc continues. "They explained to us that certain carts are 'bad' because there are a lot of big items on them." Big items had to go to shelves that could accommodate them, which were harder to find than shelves for little items. (This was at a point when the Kiva stow system had yet to be installed in Amazon's fulfillment centers.) Now that the problem was apparent, the solution was straightforward. Goods were sorted onto carts for small, medium, and large items, with fewer things going onto carts meant for larger items. Every cart became one that would require thirty minutes to stow, and one more step in the process became regular and predictable, characteristics that are essential to consistently getting things through the warehouse and always meeting the "customer promise."

This kind of process is known in lean as "standard work." Marc takes pains to explain that standard work "doesn't mean that the worker, the operator, is a machine. Standard work is defined as 'the right way to do something today.' Nothing stops you from improving that and making it even better."

Of course, this is only true for associates if management finds ways to keep them involved in that process of continuous improvement. If Bezosism were boiled down to a single thing, it's the addition of ever more automation to its own systems. This leads to ever more systemization and routinization of the work of the majority of the people who are a part of these systems.

This is where kaizen becomes a uniquely important feature of

Bezosism, the third leg, alongside Taylorism and Fordism, of the management philosophy stool upon which Bezosism rests. The key is that kaizen makes all these human-driven processes so regular and predictable that they can then be automated. Predictability leads to efficiency, which leads to more machines, which leads to a faster pace of work, which requires more kaizens in order to figure out how to make things predictable again, which just leads to more automation.

It's a cycle that feeds on itself, one which never ends. It's what observers of Amazon and many other tech giants call a "flywheel." Usually a flywheel is defined as a process that allows a company's revenue-generating engine to spin ever faster, throwing off more and more cash, which can then be plowed back into the business in ways that perpetuate its success. In this case, the combination of kaizen and automation creates a flywheel of productivity.

To return to Marc's example, the elementary process of making stowing more efficient was a precondition for the armies of robots Amazon relies on now. Bringing in the Kiva robots meant Amazon could "replace all that walking, and instead it's the shelf that comes to the worker," says Marc. "You could not have done robotization if you had not first standardized the stowing process."

"Surveillance capitalism" is a term used by social scientist Shoshana Zuboff to describe the business models of companies like Facebook and Google that make money by harvesting our data. But it's hard not to look at the state of the modern workplace as exemplified by Amazon, where surveillance is far more intrusive, and not see Bezosism as the real surveillance capitalism.

Perhaps it's because surveillance at work has such a long history, and is part of the explicit compact so many of us make with our employers, that it hasn't earned quite the same level of agita as issues of privacy and personal data have among the chattering classes. The deep roots of that compact, intertwined as they are with all the good and bad things that flow from an ever-rising level of worker productivity, are also perhaps why so many workers find themselves fighting the same battle today that their great-grandparents, or even their great-great-grandparents, did at the dawn of the efficiency movement and Taylorism.

Bezosism is a more perfect version of the time and motion studies of Taylorism, plus the automation of mass production, plus techniques for continuous improvement lifted from kaizen, plus AI and robots that can work alongside humans. But it's also very much a continuation of the struggle to balance the needs of millions of humans against the demands of the system that employs them. As Paul put it in his study of the last round of automation and intensification of work, in the 1980s and 1990s in manufacturing: "This view of Taylorism suggests that we should see Taylorism as a kind of organizational technology, and like equipment technology, it can be designed and implemented so as to empower or to enslave."

Marc Onetto has a similar perspective. "The way you use lean to motivate people is very important," he says. "In many traditional companies, you'd use this metric to reward the good [workers] and punish the bad ones. In lean, the way you reward workers is that you reward the ones who get involved in the continuous improvement work. You don't use the metric to punish people. People have a tendency to think lean is just a way to drive people crazy and get more out of them. But no, not at all. It's a way to make the job fairer and safer, and now people can perform well and be rewarded for improving the way we work."

It's a fine sentiment. It's no doubt true for some associates at Amazon. It does not in any way appear to reflect the experience of the majority of them.

Chapter 18

HOW WAREHOUSE
WORK INJURES

The effects of automation are amply demonstrated by what are by now innumerable anecdotal accounts of former employees. Injury reports from within Amazon's warehouses back up their accounts, even if Amazon can plausibly argue that these self-reported numbers may not allow for an apples-to-apples comparison with their competitors.

But what about those other effects of warehouse work that aren't accounted for in Occupational Safety and Health Administration (OSHA) statistics, such as repetitive stress injuries that might not show up on reports intended to track acute injuries, or the psychological impact of this kind of work?

Beth Gutelius and Nik Theodore, academics at the University of Illinois, Chicago, whose research was commissioned by the University of California, Berkeley's Center for Labor Research and Education, published in late 2019 the first broad study summing up all the research that has been done on the effects of automation on workers at Amazon and e-commerce in general.

"I think we're now seeing the crashing together of these old ideas about efficiency and managing workers with these new tools for doing so, which really allow it to scale in a way we've never seen before," says Beth. This brave new world of Bezosism manifests itself in two ways that are a break from the past.

The first is the effects of management by algorithm on workers' bodies. "At Amazon, it's a combination of the stress and anxiety produced by the high productivity levels coupled with the threat of being fired by an algorithm," says Beth. "It seems to me that right now when we don't have the ability of workers to push back, we have companies really testing the limits of human bodies, and the long-term impacts are unclear."

"With repetitive stress injuries at Amazon, I picture it like the housing bubble," says Emily Guendelsberger. "The profits are privatized but the risk is public, because nobody is ever going to hold these companies responsible for repetitive stress injuries that happened ten years ago. It's as if these companies are just driving a rental car and doing burnouts because they know somebody else is going to have to pay for it."

The second issue Beth discovered in her research is the profound isolation this sort of work leads to. "It really reduces the ability of workers to interact with each other," she says. "If there ever was a water cooler in this industry, it's gone now."

In words that could have come straight from the mouths of the members of Congress who grilled Frederick Taylor for three days in January 1912 about his application of scientific management to both private firms and the Watertown Arsenal, Beth found that while there is "the potential for these technologies to reduce the strain on workers' bodies, the thing is that when humans make decisions about how to implement these technologies, they are pretty focused at this point on efficiency."

The result, she adds, is that "any kind of improvements workers might gain from these technologies end up being counteracted in the way they're implemented. Now that the worker isn't doing that walking, they can focus their time on picking. So you have those gains counteracted by work speedups and a pickup in the pace of work."

The same time and motion studies Taylor and his disciples used are just beginning to be augmented by new varieties of sensors, leading to what Beth describes as the "next phase of digital Taylorism": "It's micromanagement at a much different scale than we've ever seen."

The demographic composition of this workforce can tell us a few things about who can survive in this industry, and who bears the brunt of its harms.

People who work in e-commerce fulfillment are disproportionately young and nonwhite. Twenty-six percent of workers in this sector are Black, compared with 12 percent in the U.S. workforce overall, according to data from the U.S. Bureau of Labor Statistics. Hispanic, Asian, and "other" workers are represented in e-commerce in percentages somewhat higher than in the U.S. workforce as a whole, and the proportion of white workers in e-commerce fulfillment is about 29 percent less than their representation in the workforce as a whole.

People age thirty-four and below are 64 percent of those working on the front lines of this industry, and a striking 38 percent of all e-commerce fulfillment workers are between the ages of eighteen and twenty-four. This means young millennials and members of Generation Z are represented in the e-commerce fulfillment workforce at a rate 240 percent higher than in the overall economy.

Given the overwhelming footprint of Amazon in e-commerce—the company has nearly half the entire market in the United States—it's almost certain these numbers hold true for the company. The relative youth of its workforce might indicate a lot of things. It's certainly consistent with what labor economists know about the historically terrible job market for young people without a college degree. A May 2019 report from the Brookings Institution called the market "bleak," and that was before the pandemic.

Combined with reports of injuries that, anecdotally at least, appear to be concentrated in older workers, these figures are also consistent with a feature of Bezosism that the company has never addressed directly, and one that its leaders may not have reckoned with themselves: despite its talk of automation easing the burdens of its workers, Amazon is a company determined to grab only the most able-bodied members of America's workforce. The company's relentless measurement, drive for efficiency, loose hiring standards, and moving targets for hourly rates are the perfect system for ingesting as many people as possible and discarding all but the most physically fit.

"Everyone thinks corporate cultures need to attract talent, but it also needs to repel the wrong kind of talent," says former Amazon executive John Rossman, who launched the company's Marketplace business. (Marketplace is now responsible for the majority of Amazon's e-commerce sales and revenue.)

Amazon's is a culture that's not for everybody. Driven by metrics, it's demanding and exhausting, he adds. It's no mistake that the domain Relentless.com still redirects to Amazon.com. "Relentless" was one of the names Jeff Bezos once contemplated for his then-nascent business.

John says you can compare Amazon to the Marines, or a sports team. "People try to say it's a family, but no, it's a team, and a team is performance-based. If you're not the right person to perform, they're going to trade you or cut you. On the one hand, you could call it cut-throat; on the other hand, that's what it takes to be a high-performing team."

The way this corporate culture translates to associates in ware-houses is that they're held to the same high expectations as managers, engineers, and executives.

"They are always going to be tightening down the screws on how do we scale that fulfillment process better, how do we create more flexibility, how do we lower the cost of that process," John continues. "I think that category of work and labor [in warehouses] is always prone to these pressures, and not having much leverage, these employees typically are not going to be able to have the types of environment they want. That's the constant pressure of costs and outcomes versus worker suitability. I don't know exactly where the balance should come out, but Amazon is not unique in that at all. To some degree, that is the story of the industrial and information revolution we've been undergoing for 200 years."

As Amazon itself has indicated, offering higher wages and better benefits isn't just a way for Jeff Bezos to share his wealth or burnish the company's reputation with the public. It's also a way to guarantee that Amazon has a steady supply of hundreds of thousands of the toughest and most motivated workers it can get. It's yet another way that turnover at the company is potentially to its benefit.

During my most recent visit to an Amazon fulfillment center, one of the company's public relations flacks said, "When there are criticisms [of working conditions] from outside, and from our former employees, it doesn't line up with the majority of the experiences of our associates. Our associates are our number one recruiter."

It's entirely possible, as we've seen, that both of these things are true: First, that most of the associates who have worked at Amazon for more than a year or two find the work tolerable, or at least worth the wage they're paid for it. And second, that Amazon is an extraordinarily difficult place to work, a place that sometimes does permanent damage to the health of some of its associates.

"We absolutely are proud that it's hard work here," Beth Galetti, head of human resources at Amazon, told *Fast Company* in 2019. Whether Amazon should be doing a better job of winnowing its candidate pool before they get to the point that some of them are injured is a question with an obvious answer. It would certainly mean changing the way it hires for these roles. And given its rapid growth and enormous demand for seasonal employees, it seems certain to impact the company's ability to deliver on its "customer promise."

Researchers who study the effects of laboring in warehouses warn that some of the disorders that can develop as a result of this kind of strenuous and repetitive work may not show up for years, even decades. So-called musculoskeletal disorders (MSDs) include a grab bag of the kinds of injuries that occur when a body is overused, especially in repetitive tasks. They include everything from carpal tunnel syndrome to muscle sprains, low back pain, and ruptured disks. MSDs represent 30 percent of all work-related injuries resulting in days away from work, and in 2018, 270,000 such injuries were reported in the United States. One in two American adults has some form of MSD.

Little is known about the incidence or impacts of these kinds of injuries in warehouses beyond broad statistics gathered by the federal government. We do know that the transportation and warehousing sector is the second most dangerous category of job after agriculture, forestry, and hunting. Warehouse workers in particular are injured about twice as often as the average for all workers in private industry. It's a job more dangerous than coal mining and worse than most of manufacturing. Nearly one in twenty people working in the occupations most common at Amazon will be injured or fall ill in any given year.

"Unfortunately, in many manufacturing and logistics operations, ergonomics was not considered particularly important," says Ashis Banerjee, an assistant professor of engineering at the University of Washington. That started to change in the early 2010s, with the introduction to manufacturing of ergonomics techniques pioneered decades before in construction. These techniques have yet to become widespread in warehousing and logistics. That's why, along with a team of his colleagues, Ashis is working on new, camera-based systems that use AI to watch warehouse workers and alert their bosses to ways in which they're working that might lead to injury.

Even these techniques are limited by the assessment tool they are based on, however. While researchers know what poses a human body shouldn't be in for any length of time, the gold standard for these evaluations, Rapid Entire Body Assessment, does not take into account the duration of a task or how long workers have to recover after performing it.

"Nowadays, we have lots and lots of people working in these large warehouses, and the challenge is these MSDs don't immediately show up," says Ashis. "Someone may have worked there twelve years ago, and it's not showing up until now."

Research like Ashis's begs the question: Does anyone at Amazon think about the fact that the company might be injuring its workers in ways that won't be apparent until long after they've left the company? It's clear that someone at Amazon at least suspects this is the case, because Ashis's research on how to use automated systems to alert workers to potentially injurious situations was initially funded by a grant from Amazon Robotics.

"We used to have regular monthly meetings with [people from

Amazon Robotics], to tailor our work to what they thought would be more useful to them from a practical standpoint," says Ashis.

The goal of such collaborations is, ultimately, figuring out how to make robots and automation operate in ways that are less likely to injure humans. Amazon's leaders have said repeatedly that safety is a high priority at the company.

"We have to have a high-quality work and safety environment where people enjoy coming to work every day," says Brad Porter, former head of Amazon Robotics. "Not that everyone in the world enjoys that type of work, but we're super cognizant of associate satisfaction and making the process work for them.

"It's an operating fulfillment center with associates, and engineers are working alongside them to find out what they like, what do they not like," continues Brad. "There are various mechanisms we use where we can get associate feedback, including kaizen processes where we're involving associates directly in the design and evolution of these processes. In the same way customer focus is so ingrained in how we think about things, making sure the workforce we have doing all this work is happy coming into work every day is equally a part of everything we do."

Without better reporting on injuries in warehouses, it's difficult to evaluate the outcome of Amazon's efforts to protect its associates. The current standard, which is self-reporting by companies that have no obligation to release data across their entire workforce, is clearly inadequate. On top of that is the unique challenge of repetitive stress injuries that might not show up for a long time. By the time any long-term study of their effects could be completed, Amazon could have iterated through dozens of other modes of work for the people it employs.

Even with all that data, plus a research program of appropriate ambition, there's the problem of who takes responsibility for injuries accrued years before. It's an especially acute issue in a country in which health care is employer-provided. This is hardly an Amazon-only problem. OSHA notes that for the 3 million Americans who are seriously injured on the job every year, "the financial and social impacts of [workplace] injuries and illnesses are huge, with workers and their families and taxpayer-supported programs paying most of the costs." The Government Accountability Office reports that workers and their bosses habitually underreport injuries, so 3 million is just the lower bound. Workers who experience a serious injury earn on average 15 percent less over the following decade, and they bear 50 percent of the direct costs that result from an injury. Workers' compensation covers just 21 percent.

Chapter 19

AMAZON'S EMPLOYEE-LITE ENDGAME

I found the robot I'd been searching for in a two-story warehouse in Jonestown, Pennsylvania. The air smelled like machine oil. Totes rolling down miles of automatic conveyors clattered like cicadas on a summer evening, so loud that my guide and I had to shout to be heard. Though it was daylight outside, in here islands of stark illumination pushed back a twilight made dimmer by worn surfaces, well-used equipment, and concrete floors darkened by decades of use.

Darin Russell is senior director of engineering at MSC Industrial Supply, a company that has been taking orders online and getting them into the hands of customers the next day since before Amazon Prime was even a twinkle in Jeff Bezos's eye. MSC's customers range from machine shops to many of America's largest factories. They're the sort of places that make America's factories run. When they need an item, something critical to an assembly line perhaps, they often need it as soon as humanly possible.

Darin is leading me through a maze of conveyors—here he pauses to unlatch a three-foot section of one and lift it out of our path, like a raised drawbridge—to the heart of a fulfillment center that serves the whole of the U.S. Northeast.

We pass people walking up and down aisles between walls of

shelving stuffed, floor to ceiling, with blue bins. We pause to watch one employee, a middle-aged woman, work. She uses a bar code scanner to light up the machine-readable labels on the front of a cubby, plucks an item, then drops it in a plastic tote riding on a cart. Periodically, she wheels those carts to a nearby conveyor and places the blue tote into a stream of others, all flowing in the same direction. At the end of the conveyor, they disappear into some other part of the Byzantine clockwork that surrounds us. Then she repeats the process.

"It's very simple, and very repetitive. It's really just scan, scan, scan, scan—done," says Darin. Like every other worker I'll see in my hourlong tour of this warehouse, the woman we're watching is so focused on her task that she never looks our way. We're standing maybe twenty feet from her, but it's like we're ghosts. "That's what she does all day," Darin adds.

I've come to the heart of "fulfillment center alley" to see the future. The way other people seek out rare birds, or large trees, I've come to see the thing, or at least an early prototype of the thing, that could someday transform every warehouse on Earth into a "lights-out" facility.

That's not to say the hundreds of thousands of warehouse workers currently in, for example, Amazon's fulfillment centers, won't eventually find work elsewhere. In an economy providing new services, like next-day delivery, automating one part of that service simply pushes workers toward other roles necessary to achieve it. But the robot I've come for, and the systems that support it, are, like the automated loom or the steam drill, a singular example of the kind of technological change that drives the economic "gale of creative destruction" first described by economist Joseph Schumpeter in 1942.

In the center of this warehouse, surrounded by concentric rings of automation, the robot I've come for doesn't look like much: just your run of the mill industrial robot arm, pneumatic tubes snaking down its length, ending in a suction device for grabbing items. Darin fires it up, and the hiss and pop of its air-powered system for grabbing packages adds to what was already a loud din.

On one side of the robot arm, the same totes filled by the workers we observed earlier travel down an inclined conveyor. One at a time, a tote slides into the robot's domain, a chamber about the size of a refrigerator, its walls made of plexiglass and steel. A camera looking down on the tote captures an image of its contents, and in less than a second, an AI examines that image and develops a hypothesis about how to grab the item. The multijointed arm then swings into the tote like an elephant's trunk, its pneumatics hissing and popping.

Sometimes—often, on the day I visited—the arm makes a mistake: It can't grab the item because of the shape of the squishy, slippery poly bag in which it's been packed. Or a bunch of small items, boxes of screws, say, have been bound together with a rubber band by a far more dexterous human, and now some have come loose. The arm might try again, but sooner or later, it knows it's been bested by the endless variety of the material world and calls for help.

In San Antonio, Texas, 1,500 miles away, an employee of Plus One Robotics, the company that sells the system used by MSC, gets an alert, and the same image captured by the robot's cameras pops up on their screen. Unless there's a waiting list of requests from other flailing robot arms overseen by Plus One, the employee addresses the remote arm's request immediately. On their own screen, they use a mouse to click on the center of the item in the bin. Then they click on a few icons to pick the correct gripper, and to give the remote robot arm a hint about how to pick it up, and the arm swings back into action. Usually, it can now pick up the item successfully. Next, the robot drops the item in an appropriately sized cardboard shipping box, produced entirely automatically by a different machine. Finally, the robot pushes the completed order onto a nearby conveyor. Once the box is closed and sealed by yet another machine, it's ready to be packed into a truck and shipped to a customer.

This entire system, and all of the decades of research into robotics, artificial intelligence, and machine vision it embodies, is here just to replace a single human being taking items out of totes and consolidating them into a box.

At one point, an item gets stuck in a corner of a tote as it comes into the robot's area of action, and Darin has to open the gate that separates us from the robot, which makes it retract and shut down, for our safety. Using his vastly superior system of object recognition and manipulation, Darin places the item flat in the center of the tote so the robot can grasp it. This happens with robots of this sort. At MSC, 95 percent of the time, the robot can handle things on its own. Most of the rest of the time, a remote human can help guide it. And perhaps one time out of a hundred, a person who is physically present must step in to solve the problem on-site.

Today, one person in a warehouse can oversee twenty robots, says Erik Nieves, Plus One's animated, voluble founder. "But the same crew chief handling eight robots today is handling twenty robots tomorrow, and we've run models that show forty to fifty robots per person," he adds.

Driving those potential gains in efficiency is the feedback loop between the human minders in San Antonio; the users of these robots in warehouses, which include not just MSC but FedEx and other Fortune 500 companies; and the AI that powers them. Every time a robot fails and a human corrects it, it can learn from its mistake.

That doesn't mean humans will ever be out of the loop completely. "I'm not an AI skeptic, I'm just a human exceptionalist," says Erik, who was an executive at Japanese industrial robot arm manufacturer Yaskawa before he founded Plus One. He believes that, given the unending variety of goods stocked by a company like MSC, recognizing and picking up all of them will always require humans, both remote and present.

MSC stocks more than 800,000 different types of items, says Doug Jones, the company's chief supply chain officer. "The challenge has been putting that level of complexity in front of a robotic arm and asking it to detect and pick up all these individual items," he adds.

Despite relying heavily on automation elsewhere in its supply chain, MSC hadn't previously added any robot arms because they couldn't handle its breadth of goods. It's the same reason that Amazon, despite massive investment in robotics research and development, still relies on humans to stow, pick, and transfer goods throughout its warehouses.

The first time MSC tried one of Plus One's robots on its distribution line, the robot successfully picked and transferred between 60 and 70 percent of items, says Doug. "It doesn't sound high, but it blew us away," he tells me. Now the success rate for the robot arm is above 95 percent, and almost all the rest of the cases it encounters can be handled remotely by Plus One's humans.

MSC's leaders are unabashed about getting humans out of the company's supply chain. The paradox of warehouse jobs, even in an economy in which many low-wage workers have been laid off, is that it's hard to hire people willing to do them for any length of time, says Doug. They are the classic dull, dirty, and difficult jobs where turnover is high and physical fitness is not optional.

Plus, he adds, everyone expanding their e-commerce operations wants to put their fulfillment centers in roughly the same locations. They need to be on major arteries—MSC's Jonestown warehouse is at the intersection of I-81 and I-78—and close to downstream UPS and FedEx hubs that encompass a zone from which deliveries can be made overnight by truck. They've also got to be in places where there's enough land to accommodate them.

Rural and exurban areas where distribution and fulfillment centers are rapidly proliferating, including Fulfillment Center Alley in Pennsylvania and the area around Reno, Nevada, don't have enough workers to fill all their open positions. As warehouses continue to sprout, everyone in these areas increases wages in order to attract workers. In the short term, this is good for workers, but in the long run, any gap there once was between the cost of a machine and the wages of a human grows narrower.

"The beauty about a robot arm is, it doesn't call in sick, it doesn't take breaks, it doesn't need light," says Doug. Without queuing time or idle time, it also makes other investments in MSC's warehouse more effective—such as the entirely automated packing, sealing, and labeling machine downstream of the robot arm. That machine is itself a wonder of automation, but hardly a new one. By now mainstream throughout Europe, many companies make such devices, including CMC, which has sold machines to Amazon.

The package we saw the robot arm pack travels down yet more conveyor, until it arrives at the back of a truck, where it will be loaded by a human. With the addition of Plus One's robot arm, no human touches an item from the moment it's picked until the moment it goes onto a truck, says Doug, and even those steps might someday soon be automated, he adds.

Fully automating most every other job in the warehouse is the ultimate goal of technologists building robotics of the sort being developed by Plus One and its many competitors, including Covariant, Osaro, Dexterity Robotics, Pickle, XYZ Robotics, Mujin, Dorabot, Fizyr, Soft Robotics, Kindred, and RightHand Robotics.

While systems from companies like Plus One can transfer a wide variety of items from a tote or a conveyor, getting them out of the kinds of densely packed storage that is common and necessary in most warehouses will require quite a bit more innovation. Jeff Bezos said in 2019, "I think grasping is going to be a solved problem in the next ten years. It's turned out to be an incredibly difficult problem, probably in part because we're starting to solve it with machine vision, so machine vision did have to come first."

Bezos should know: Amazon has invested billions in automating its logistics and is continuing to hire and cultivate the kind of talent required to create entirely new forms of automation. In addition to work at its Seattle headquarters, the company announced in 2019 that by 2021 it would open a 350,000-square-foot robotics innovation hub in Westborough, Massachusetts, near Boston.

If Bezos is right, the implications are tremendous: Amazon could replace the overwhelming majority of its permanent workforce— hundreds of thousands of associates—with robots. In order to compete, every other company in e-commerce would have to follow suit, which would mean millions more workers would be shunted out of warehouse jobs. As quickly as the pandemic pushed low-wage workers out of jobs in service and retail and into the arms of the rapidly expanding e-commerce industry, in a decade or two they could be in for just as rapid a disruption as robots take over those jobs.

A number of technologies will have to converge in order to fully automate picking items out of totes or off shelves. Some of these challenges are relatively easy to solve. The majority of industrial robots sold today still go into automobile manufacturing, which requires extreme strength, durability, and precision, all of which come at a high cost. Making robot arms less of all those things, and a good deal cheaper, will be essential to making them mainstream in logistics, but that's just engineering. The real challenge is turning the suction device or gripper at the end of that arm into something as flexible and dexterous as the human hand.

Companies like Soft Robotics have created flexible hands made of silicone—the same super-durable stuff that goes into bakeware—that are powered by air. Soft Robotics' technology was initially developed by George Whitesides, a chemist and materials scientist. Deriving inspiration from the mechanics of an octopus tentacle, his team won a 2007 challenge from the Defense Advanced Research Projects Agency to create a robot that could slither under a door. Theirs resembled a starfish. The ultimate commercial result of the team's research wasn't an autonomous and creepy superspy, however. The team transformed their work into a gripper strong enough to pick up a gallon of milk but also delicate enough to handle the marshmallow snacks known as Peeps. Many imitators have followed.

One reason humans aren't being replaced faster in warehouses is that the explosion of e-commerce is such a recent phenomenon. Whereas manufacturing has experienced more than sixty years of compounding progress in robotic arms—the first industrial robot, Unimate, was installed at a General Motors plant in 1961—in logistics, they're practically brand new. "It's always everybody's first robot that's the issue, but unfortunately in supply chains, any robot you put in is their first robot," says Erik.

In terms of the evolution of warehouse robotics, "we're in the first inning—the first at bat, really," says Tye Brady, chief technologist at Amazon Robotics. "I'm a rocket scientist," he continues. "It may seem like worlds apart, landing on the moon and e-commerce fulfillment, but they're actually very similar."

What space exploration and next-day delivery have in common is the hybrid nature of humans and the automation required to accomplish them. One of Tye's former gigs was an eight-year stint as a spacecraft engineer at Draper Laboratory, a not-for-profit research outfit in Cambridge, Massachusetts, that is, among other things, collaborating with NASA on returning humans to the moon. Whether it's a lunar lander or a Kiva drive unit in an Amazon warehouse, getting anything done in the real world requires humans and machines working together.

"Systems engineering is about having machines and people each do what they do best," says Tye. "The end goal is to help the human do things the human cannot do alone."

Marc Onetto, the former senior vice president of worldwide operations at Amazon, doesn't believe humans will ever be eliminated from the supply chain. Just as the rise of e-commerce both destroyed millions of jobs in retail and created millions of jobs in warehousing and distribution, humans will always be needed for the next industrialized convenience offered by the next Amazon. "I always take the example that when the train came, it was bad news for the stagecoach drivers," says Marc. "Yes, we killed the stagecoach driver's job, but it was good for humanity."

At MSC, the person whose job isn't going anywhere is the "water spider," whose job it is to step in when the roots have a problem they can't solve on their own. There are also the technicians and engineers who maintain all the automation.

But it doesn't take nearly as many people to oversee robots and automation as are required to do the work that they do, and that's the point. At the MSC distribution facility, which at the time of my visit in 2019 had only one robot but was in the process of adding more, the entire staff across three shifts numbered 400, including custodians and management. Just seven of those 400 workers were maintenance technicians.

Thus, an 800,000-square-foot facility pushing out 25,000 items in 8,000 to 10,000 orders a day requires only about a hundred people per shift, says Darin. In the past few years, the number of items offered by MSC has increased, and the number of orders flowing through this building has too, but the headcount in Jonestown has not.

Without automation, MSC would not survive, says Doug. Before recent waves of automation, it took the company three hours, on average, from when it received an order to when it was boxed and placed on a truck. Now, with automation, that's down to forty-five minutes. This kind of turnaround time is critical to getting orders to MSC's carrier, which is almost exclusively UPS, in time for customers to receive them the next morning.

The short turnaround time for an order is also an example of the way that the efficiencies of automation can beget yet more efficiencies. The faster an order moves through the systems of a warehouse, says Doug, the smaller the warehouse can be, because its miles of conveyor must hold that many fewer in-process orders. Compared to a system in which it takes several hours for an order to wend its way through a building, one that can do it in forty-five minutes needs only 20 percent as much floor space devoted to order processing. That, in turn, allows a company to squeeze more and more life out of its existing facilities, even as the volume of orders grows year after year.

Erik Nieves doesn't believe that we'll ever get to the point that warehouses are completely lights-out, for the simple reason that the seasonal fluctuations in shipping volume will always necessitate hiring temporary workers to account for floods of orders.

"You have this set of people out there who believe you're going to automate the whole thing and it's just going to run in the dark, but that requires building the church for Easter Sunday," Erik says. Over the holidays, or during a sales event like Amazon's Prime Day, the volume of goods flowing through a distribution center can spike to two or three times the volume the facility would normally handle. The flexibility of humans means that keeping some around, no matter how automated a facility, allows companies to rotate employees through whatever job they're most needed for at any given time.

There is also, of course, the possibility that human labor becomes cheap again. "My existential threat as a robotics company is comprehensive immigration reform with a guest worker program," says Erik. "But we're further away from immigration reform than ever in this country, so what is a labor-constrained operator to do?"

For now, across the entire e-commerce industry, the answer is hire, hire, and hire some more. The boom in employment in e-commerce has meant that the industry has, as a whole, created more jobs than have been lost in physical retail, at least from the period of around 2007 until 2017.

Beginning with the mobilization of America's industrial might after it declared itself a party to World War II, America saw a boom in the demand for workers in factories. Today, the number of people in America employed in manufacturing is the same as it was in 1941, when only a third as many people called the United States home. It's as if eighty years of global manufacturing dominance never happened. In terms of the number of people employed in the business of making things with their hands, we are back to where we were at the end of the Great Depression.

Much of that is due to outsourcing, some to automation, but the two are not mutually exclusive. As warehouses and the rest of the supply chain become ever more automated, it's hard to imagine logistics won't eventually follow the same trajectory as manufacturing.

Chapter 20

THE MIDDLE MILE

One of the mind-boggling things about America's fulfillment and distribution networks is that sometimes it makes sense to put even fairly common and widely distributed items on a truck and drive them clear across the country. It is, one supposes, no different than the fact that it makes economic sense to produce an object that costs less than a dollar in a factory that is geographically the farthest possible point on Earth from that object's ultimate destination.

This is how the box containing our USB charger finds itself on a second, thousand-mile journey over America's highways, to a facility known as a sortation center. This journey, from the fulfillment center, through one or more sortation centers, to a delivery station where a package will finally go onto the truck or van or (someday) drone that is to deliver it to your house, is what's known in logistics as the "middle mile."

Few people outside of the logistics industry talk about the middle mile. Like every second episode of a trilogy, it has the distinction of being mostly connective tissue. And yet, taken as a whole, the middle mile is by some measures the biggest, fastest-moving, and most fully automated portion of the entire journey of a package.

If an Amazon fulfillment center is something like a digestive system, breaking down pallets of goods and "individuating" them into packages before sending them on their way, then what comes next is a sort of circulatory system. It's charged, through its endlessly branching byways, with getting goods to their final destination—your home.

These networks are older and, collectively, bigger than those of even mighty Amazon. UPS, founded in 1907, has been building its network for over a hundred years. FedEx, while founded in 1971, has grown nearly as large. The two companies are, as of 2020, closely matched in both revenue and the scale of their respective networks. UPS has for the most part caught up to FedEx in air freight, and FedEx has mostly caught up to UPS in ground shipping. UPS is still about half again larger than FedEx by overall shipping volume, delivering 4.7 billion packages in 2020 to FedEx's 3 billion. The reason their revenues are nearly the same is that FedEx earns better margins moving more high-value overnight packages.

With their differing histories and corporate philosophies, they are hardly copies of one another. But UPS and FedEx have by now co-evolved to the point that they are like fraternal twins, with more traits in common than not. Together, they have 80 percent of the market for delivering packages in the United States. Almost the entirety of the rest of that demand is carried by the U.S. Postal Service.

It's important to qualify what I mean by "the market" here, since it excludes the largest private delivery network ever conceived—Amazon's. In June 2018, Amazon announced that entrepreneurs would have the opportunity to start their own regional delivery services, buy their own Amazon-branded vans, and start shipping for Amazon as independent contractors. Just eighteen months later, analysts predicted that Amazon would soon be moving more packages than FedEx. By the middle of the 2020s, Amazon Logistics, as it's known, is projected to take the number one spot from UPS. By one estimate, the volume of Amazon's air and ground delivery network grew 155 percent from 2019 to 2020 alone, from about 750 million to 1.9 billion packages.

Whether it's UPS, FedEx, Amazon, or the United States Postal Service (USPS), all nationwide shippers must use sortation centers to route goods as they go from fulfillment centers to the stations where they are put onto trucks and vans for final delivery. For any package that must travel more than an hour or so, it isn't practical to carry it directly from a fulfillment center to its destination on a delivery truck. So our USB charger must be grouped with other packages and taken on a semitrailer truck to a sortation center. From there it may go to an attached delivery station, where it's put onto a delivery truck, or it may go onto a second semitrailer truck that carries it to its final warehouse, a delivery station.

▶ ▶ ▶

From the outside, the 625,000-square-foot FedEx Ground sortation center in Hagerstown, Maryland, looks like an aircraft hangar, only longer. As I drive from one end to the other, it keeps going, and going, an illusion enhanced by the neat, endless rows of FedEx trailers in front, every one of them bright white, with a crisp purple-and-orange logo.

My guide today is Ted Dengel, managing director of operations technology and innovation for FedEx's ground-based shipping network. Where Amazon's managers and flacks talk about their employer in a way that reminds me of every tech start-up ever—that is, in a specialized and cultish language, with an emphasis on how everything they're doing is going to change the world—Ted talks about what FedEx does in the context of the company's history. He rattles off hard numbers, interrupts himself with caveats, gets a little wistful about how things used to be done, widens his eyes when emphasizing how much better they are now.

When I ask him whether the hardware inside of the facility we are touring is all that different from any other sortation center owned by his competitors, he pauses and then says no, not really. They're all trying to solve the same problems, and they're all buying automation—sorters, conveyors, scanners, tugs, and the like—from the same vendors.

What makes FedEx different, he adds, is how it uses software and, as of about 2018, artificial intelligence and machine learning, to optimize its entire network. All that high-level engineering happens in

Pittsburgh, where FedEx employs between 400 and 500 engineers and can draw on the pool of industrial engineers and computer scientists coming out of Carnegie Mellon and other nearby universities.

From the perspective of an individual package, it doesn't much matter which of America's shipping giants carries it through the middle and last miles from a fulfillment center to your front door. In the overwhelming majority of cases, what happens in this facility in Hagerstown is squarely in the main of how that process plays out for an average of 40 million packages per day, as of 2020. To put that another way: for every person in America, almost fifty packages a year are shipped, and that volume is steadily growing.

The process of sortation begins when the back of a FedEx semitrailer rolls up to a loading dock at the sortation warehouse. Doors rolled up, a FedEx employee has to unload the stuffed-to-the-ceiling trailer onto a conveyor that terminates right at the edge of the loading dock, at the boundary of where the semitrailer is mated to the building.

For the vast majority of packages that enter the building, when someone unloads them from a truck, that's the last time a person will touch them until they are packed onto a truck by another human, at the other end of the building. The average sortation center is therefore the most automated waypoint in the entire journey of our USB charger. This is especially notable in light of the fact that the automation in all but a handful of the most innovative sortation centers hasn't changed much in the past decade or two.

In some ways, this level of automation is possible because sorting packages isn't nearly as hard as storing, retrieving, and boxing up individual consumer goods, which is what fulfillment centers must do. Once they arrive here, they may come in every imaginable shape, size, and variety of packaging, but at least they're all parcels of one sort or another.

In the fictionalized but plausible chronology of our USB charger, it's March 22, two days after it left the fulfillment center in Minnesota. Most items originating from a fulfillment center in the Midwest stay in that time zone. But some keep going, all the way through the Rust Belt to the East Coast. Here, for the sake of both narrative continuity and journalistic breadth, I've taken this liberty: No Amazon package would ever go through a FedEx sortation center. Instead, they would go to one of Amazon's own, or one run by UPS or USPS.

As our USB charger has come so far since the start of its journey, so has the world. When it was on the quayside in Vietnam, the novel coronavirus was just making headlines outside of China. By the time the charger would have made it to a person's home, they'd be well into their first lockdown, ordering everything online. On March 22, the mayor of Washington, D.C., announced she'd use the National Guard to keep crowds from gathering to see the city's famous cherry blossoms; Philadelphia declared a stay-at-home order; California and Washington state both declared a "major disaster."

The three-story FedEx sortation center in Hagerstown is by far the loudest industrial facility I've ever been inside. Once I'm through security and inside the churning heart of the building itself, we're shouting at each other to be heard. The situation isn't helped by everyone's masks, a necessity on the day of my visit, in mid-October 2020, when Covid-19 cases are spiking across the United States yet again.

One reason for the noise is that this facility is running all-out. The day before my visit, the volume of packages flowing through the geographic area served by this sortation facility matched the busiest day of the holiday season the year before. And the holidays are of course inevitably the busiest time of all for shippers in the United States.

At the moment of my visit, the nationwide shift from going into a store to buying even basic necessities online was so widespread, so profound, that the busiest, most infrastructure-straining day of a typical year for FedEx, UPS, and Amazon was now their new normal.

The only reason FedEx has been able to cope, says Ted, is that its facilities were built to handle that much capacity, as long as enough people can be hired to make everything run at its maximum rate. Raising starting wages for package handlers at FedEx to $18 an hour at this facility, in an economy devastated for retail, food service, and hospitality workers; extending shifts; and adding overtime did the trick. FedEx Ground also moved to a seven-day-a-week delivery schedule in 2019, a move that proved prescient.

Despite being fully automated in its primary functions, this sortation center employs a total of 1,400 people, some of whom are part-time. In addition to unloading and loading semitrailers at either end of the warehouse, most of the people in the building are handling items too large or too small for automated sortation. Even though most of the parcels flowing through this facility are of the medium-size variety—larger than a single T-shirt in a poly bag or document in a flat mailer, smaller than a big-screen TV—all those items that aren't in the middle of the bell curve distribution of package size require many human touches in order to flow through the building. It's another illustration of the way in which automation in logistics still requires a great deal of human involvement in order to handle functions machines can't.

At its most basic level, the purpose of this sortation facility is to take packages from semitrailers plugged into its 56 bays and move them as quickly as possible to one of 240 end points, each a conveyor feeding a different truck. From the moment a package leaves one truck to the moment it arrives at the end of a conveyor to be placed on another, the entire process happens in, on average, three and a half minutes.

Its relentless pace is the other reason the inside of this building is filled with a sound like the whine of a jet engine at full throttle on takeoff. Packages zip down conveyors moving at just over six miles an hour, which might not sound like much but is plenty impressive when you're standing right next to such a conveyor, filled from end to end with parcels.

My tour begins at the "front" of the building, where the whole process starts. A long section of warehouse about 150 feet across, it has twenty-eight truck bays on either side, open to the outside world like spiracles on the thorax of some gigantic insect. At maximum capacity, a person unloads packages from the back of a semitrailer at every one of these bays. The moment after that person pivots their body and places a package onto the motorized conveyor directly behind them, it climbs toward the ceiling at a 45-degree angle, up fifteen feet into the air, where it's shunted onto a wide conveyor that carries it down the length of the building.

As soon as the package has reached the end of the portion of the building that includes truck bays, the building widens, and the package hits another conveyor that forces it to make a hard left turn. As Ted and I stand to one side of this conveyor, fifty feet down from this turn, we watch packages move—dance, really—in ways that feel strangely animated. It's almost as if we're watching race car drivers sort themselves out while going into a turn, each jockeying for position. Some packages speed up, others slow down; some pirouette as sections of conveyor move at different speeds, in order to space out packages in a process called "singulation." The system uses cameras to watch packages shoot by, accelerating or throttling the speed of different sections of conveyor, figuring out exactly how to tweak the endless stream of parcels so it can be ready for the next step.

Packages that are too crowded with their neighbors, or that aren't oriented correctly as they roll down the conveyor, are automatically pushed sideways, off the main conveyor, down a small slide, and onto a return conveyor that feeds them back into the primary stream of other parcels. This sort of physical error correction is common in automated warehouses. For example, it's an essential part of the Amazon Fulfillment Engine inside its sortation centers. It's the physical-world equivalent of asking someone to repeat themselves in a conversation, or the automated error correction and resending of packets of information built into the internet and most any other sort of electronic communications network.

After singulation, every box and poly bag that goes through this facility, save the very largest and smallest ones, speeds through a tunnel of red light. Here, a camera system takes pictures of the box on all six sides. These images are stored locally, in industrial PCs attached directly to the scanning system, so that FedEx can retrieve a complete image of any package that flows through this building, should there later be any question of when or how a package went missing or was damaged.

Then the packages take a hard right turn onto one of eight "secondary" conveyors, and this is when the business of sorting them really begins. In FedEx's facilities, this is primarily accomplished with a device called a "shoe sorter," one of those little-known but essential and ubiquitous workhorses of the logistics world.

If every automated facility in a logistics chain were a microchip, its conveyors would be the etched channels carrying data, in the form of electrons, from one place to another. In this analogy, shoe sorters would be the transistors of the chip, the parts that perform binary logic on those streams of data-as-electrons. And just as everything about our modern digitized world would be impossible without transistors, nothing in the heavily automated world of logistics would be possible without some sort of mechanical device capable of deciding, at high speed, whether a package, like a bit, should be diverted to one path or another—should be rendered into a one or a zero.

History is full of devices capable of performing this function, which was once solely the domain of humans, and there are more complex ways to sort items—say, with robot arms. But for the highest-speed operations, up to 220 cartons per minute, or one every 0.27 seconds, shoe sorters are the most common. They are the most basic, most widely used, most essential device in an automated warehouse. They are the least dispensable constituent parts of what are essentially giant mechanical computers acting on atoms instead of bits, and FedEx has been using them for close to twenty years.

In their operation, shoe sorters are deceptively simple: they're just conveyors made of slats, with small bumpers at one end of each slat. As a shoe sorter whizzes by a fork in the conveyance system, a branch pitched at around 22 degrees for the highest-speed systems, the slats and their attached bumpers move perpendicular to the direction of the conveyor, gently sliding a parcel off the main path and onto a different one.

Put enough shoe sorters in a row, and you can use binary logic to sort packages with a level of complexity limited only by how many square feet of warehouse are available for conveyors.

At this facility, the single primary conveyor uses eight shoe sorters to divert packages onto one of eight secondary conveyors, which flow into the final third of the building. Each secondary conveyor runs past one of thirty gravity-fed spiral chutes that take packages from the second-floor mezzanine to the ground floor. Eight times thirty is 240, and by this simple math, conducted on a stream of packages as if each were the beads of an abacus, the boxes and bags that flow in one end of the FedEx sortation center are subdivided into 240 different end points, each of which is a conveyor that ends at the mouth of a different semitrailer.

The decisions that must be made at each of hundreds of branches in the system are made by a symphony of individual computerized components, all coordinated by a central system—a single conductor, sitting in the control room of the facility—and above that, by an even larger system plotting and directing the movement of millions of packages a day.

In the building's control room, it takes only three people to direct the operations of the whole building. One paces before a bank of six screens. The three screens on top are subdivided into a grid of feeds from cameras all over the facility. From these, the facility director can view any branch in the sortation center and any of its operative machine or human bits. The bottom three screens are all cryptic with dense information. In one corner, a bar graph updates every second or so with the total throughput of the facility—more than 20,000 packages an hour at the moment I glance at it. The rest of the bottom screens are taken up with representations of every chute in the facility, gray for all clear, orange if a package has gotten stuck in one and is impeding its throughput, red if it's become blocked or is completely full.

In addition to the fully automated portion of the sortation center, these screens also represent the parts that still require many hands to convey and sort packages. On the first floor, where we head next, long chains of rolling carts are driven by people in small wheeled tugs who must perform the same role as the automated secondary conveyors above them, only they are transporting the extra-large packages, too big for conveyor belts.

On these carts I see mail-order mattresses from start-ups of the sort that advertise heavily on podcasts and Instagram. There are also big-screen TVs, flat-pack furniture, and tires—so many tires. Who knew FedEx is how America moved this much of the vulcanized rubber with which our more than 270 million vehicles are shod? There are also many boxes branded with the logos of a variety of other large retailers, which makes sense: Since FedEx dumped Amazon as a client, it has become a carrier of choice for many of its competitors.

In other facilities, FedEx is working on automating the transportation of these large items, says Ted. In some cases, FedEx is using bigger conveyor belts; in others, engineers are replacing the human drivers of the tugs that pull chains of carts with autonomous systems that perform the same role. It's the fruits of all the research that have gone into the autonomous-driving industry distilled into a much less expensive system that need only handle a much simpler environment.

Elsewhere on the first floor of the building, packages below a certain size, known as "smalls," are sorted the old-fashioned way. These roles are the ones that most closely resemble the tasks taken on by Amazon's stowers and pickers. And here, it appears to be a very similar story: the pace of the automation is so prodigious that, even as it eliminates some tasks and makes others easier, it turns those that remain into a sprint for the humans who remain in the role of the biological connectors who must link one automated system to the next.

Where in the past most small parcels might be letters or documents, in an age of e-signatures and e-commerce, they're mostly individually packaged consumer goods. Half of it, at least from what I can see, is in the form of soft poly bags holding one or two garments at most. It's a sea of packaged T-shirts, socks, underwear, tights, sweatpants, and all those other individual items a shopper might idly drop into their cart on a vendor's website or app.

These parcels, diverted off of conveyors and onto broad chutes, funnel down to a single person standing in the center of a floor-to-ceiling column made up of smaller chutes. Each of these workers is the hub of a three-dimensional wheel, where each spoke is ten feet tall and consists of mail slot–size openings. Every one of these sorters works as fast as they can. First they grab a package off the chute, then they pause for a moment to scan the item and read its destination off a screen; at the same time, an audio prompt conveys the same information. Then they whirl and drop the item into a slot.

Each of these workers must sort between 1,100 and 1,200 parcels per hour, says Ted. At the high end, that's twenty a minute, or one every three seconds, without pause, for a four-hour shift. As in the Amazon fulfillment center, everyone here, no matter their age, displays levels of physical strength, agility, and endurance that make it clear the average office worker would be hard-pressed to keep up.

As Americans order more and more online, in smaller quantities, this portion of FedEx's business is growing quickly. The need to automate the sorting of these small packages means the company is testing a number of different technologies, says Ted. It's been a challenge to find ones that are as good at the job as humans, but the technology is getting better fast. "Our issue is the variability in the package mix," he adds. "Because they're mostly poly bags, we don't see the productivity out of those automated solutions that we see here."

Watching the arms of the human sorters windmilling, I marvel that their job is basically a four-hour sprint. "Yeah, it really is," says Ted. Sorting small packages isn't as physically demanding as transferring heavy ones, he adds, but it requires a person to think and process information quickly, and for hours on end

At the end of all this sorting of parcels of every size, all arrive at one of those 240 conveyors, where a human must reverse the process that occurred at the start of the sortation center. Here, a loader must fill the back of a semitrailer as tightly as possible, without crushing any packages on the bottom of the trailer with excessively heavy ones on top. A system known as "Trailer Loading Analytics" measures the density of the packages they manage to load into a trailer, making sure they're filling the trailer completely, with no gaps between packages, and within a prescribed amount of time.

Once fully loaded, these semitrailers will take packages to their second-to-last destination, the yet smaller warehouses where they are put on vans and delivery trucks and carried to homes and businesses.

One such delivery station is attached to this sortation center. In the past, says Ted, FedEx and other carriers relied on a hub-and-spoke model to take packages from large central facilities through ever smaller ones, like blood flowing from the heart into arteries and, ultimately, individual capillaries. Now the model is, by necessity, quite different. The entire network is more of a "mesh," a system in which every single ground station can both sort packages and get them onto delivery vehicles. Such a system is more complicated to operate and requires more automation, but it's also a great deal more flexible. Packages can flow in both directions through such a nonhierarchical system. In this new model of distribution, what varies is the size of individual sortation and delivery facilities, more than their function.

This delivery station represents the boundary between the middle mile and the last mile. And here, our USB charger, having been touched only twice by humans in its entire trip through this vast building, begins the final leg of its journey.

Chapter 21

THE FUTURE OF
DELIVERY

It's 11:00 on a sunny, bright blue sky and stiff cold breezes kind of fall morning on the Disney World–immaculate campus of George Mason University in Fairfax, Virginia, and my guide and I are chasing a cup of Starbucks iced green tea.

The tea is ensconced in what looks like a futuristic cooler on wheels. It's actually a delivery robot, ambling down the sidewalk. Whenever it detects a pedestrian crossing its path, it freezes like a startled animal.

At one point, a young man actually steps over the two-foot-high robot, forcing it to stop so abruptly that its back kicks in the air a little. Then a young woman maneuvers around it without even looking up from her phone. As the robot comes to a particularly congested stretch of walkway, a cataract hemmed in by fences and construction on both sides, students power walking from one class to another ignore it completely, slowing its progress to a hesitant crawl.

If there is one thing humans have an astonishing capacity for, it's taking things for granted—especially new and potentially transformative technologies. "Remember the first time you used Wi-Fi?" says Ryan Tuohy, a vice president at Starship Technologies, maker of the delivery robot we're tailing. "It was pretty exciting," he continues, and I nod, because I too am old enough to have been fully sentient in the late 1990s.

I feel a bit like I'm the only person here who can see this thing, this little robot, which is maybe a ghost and I'm dead too? But I get it, and Ryan confirms it's this way everywhere Starship rolls out its robots: The first few times someone sees them, they're remarkable, a bit of the future poking into the present, adorable but also a bit unnerving, and very likely the first autonomous robot a person has ever encountered.

But eventually, it's just someone's drink order, and it's in the way.

A few minutes later, we're in a less crowded place, a paved plaza with a bronze statue of George Mason himself at its center. An alert appears on Ryan's phone, letting him know his tea has arrived. We walk up to the robot, which is secured with an electronic lock, and Ryan hits a button on the Starship app to let it know he's here, too. The lid of the robot unlocks; Ryan opens it, and inside the insulated cavity of the robot's body, there's his drink, secured in a caddy. As he reaches for it, the robot cheerfully declares, "Hello! Here's your delivery!"

The whole process seems so simple. Fifteen minutes ago, Ryan ordered his drink on the Starship app, which sent the order directly to the on-campus Starbucks. Ten minutes ago, a barista walked out of the double glass doors of that Starbucks, opened the lid of the waiting Starship robot, and placed the drink inside. Then, with only a minimum of human intervention, it drove to us, never faster than four miles an hour.

But to accomplish this sequence of events tens of thousands of times a week, in dozens of cities all over the world, in every kind of weather, and to make delivery via robot into a going concern—a real business that can eventually make money—is the real trick. While fully autonomous vehicles remain mostly a science project, as well as a potentially risky ongoing experiment on America's roads, and drone delivery awaits both a workable business model and final approval by the FAA, Starship is the one company on the planet that, as of this writing, anyway, is doing autonomous delivery at scale.

It's not as if Starship doesn't have plenty of competition, of course, with Amazon, FedEx, and countless start-ups all testing their own versions of delivery robots. Considering the scale of the industry it's tackling—globally, food delivery on its own is already a $100 billion business, and package delivery is about five times as large—it's clear there will be many companies that will eventually send their cooler-size robots onto our sidewalks and streets. Today's delivery of late-night munchies on the campus of George Mason University is to-morrow's delivery of the packages we order online. Perhaps especially those small, critical, last-minute needs.

Imagine the USB charger we tracked across more than 10,000 miles ferried on the last mile of its journey not on a delivery truck but in a robot. And instead of it taking a day or more to reach us, it takes minutes. It arrives directly at the door of the café, office, home, or co-working space where we happen to be doing something important but find we've forgotten a way to keep our device powered up.

Already, you can order a MacBook or any of a number of other electronics from Apple in many major American cities and it will arrive in two hours, courtesy of Postmates—which is also, not coinciden-tally, working on its own robotic delivery service. "We do a ton of Apple delivery; you'd be surprised," says Ali Kashani, head of the skunkworks at Postmates (now owned by Uber) working on the com-pany's autonomous robot. "A lot of it is iPhones, AirPods, charging cables. Things like these are very popular because they're the things you need right away."

Already, Starship delivers packages in Milton Keynes, a bedroom community of London. Every delivery costs £1, while at George Mason, it costs $1.99.

While my visit to George Mason comes less than a year after Starship launched on campus, there are people here who have already used the service more than a hundred times. It makes sense: the ro-bots integrate directly with vendors owned by Sodexo, the on-campus food provider, so students are already using their meal plans to order from Starbucks and a half dozen other restaurants on the grounds of the university. For a couple extra bucks, they can save themselves a trip across campus, waiting in line, transacting, all the rest. It's the same convenience they get from an Uber Eats or a Grubhub, but the delivery fee is a fraction of what they'd pay those services—not to mention that you don't have to tip a robot.

The twenty-five robots on campus the day I arrive are busy, operating at their maximum capacity of about two deliveries an hour throughout peak times, such as lunch. Ryan calls Starship's customers on campus the "on-demand generation." Members of this generation, also known as Generation Z, have never known a time when they couldn't order pretty much anything on their phone with just a few taps: A meal, a ride, the tens of millions of items on Amazon, even a hookup.

It's easy to see how two-dollar, thirty-minute deliveries could turn us all into members of the on-demand generation, but first Starship has to grow its business, one delivery area at a time. College campuses are a good fit for its initial rollout, since they concentrate a sizable, hungry, and busy population into a small, pedestrian- and robot-friendly area.

A college or corporate campus—Starship also operates in the office parks of Intuit, Volkswagen, and Bayer—is not the only place the company's model works, of course. When the pandemic hit, not long after my visit to George Mason, Starship distinguished itself as one of only a handful of autonomous delivery companies in the world to actually expand its service.

After George Mason decided to allow many of its international and graduate students to quarantine on campus, Starship's robots continued to deliver, even during local and state lockdowns. Then the company expanded its delivery area to neighborhoods adjacent to campus. In the early months of the pandemic, Starship also rapidly stood up new delivery services not attached to any college at all, for example in Tempe, Arizona.

In all the dozens of locales it operates in worldwide, Starship, like all delivery services, saw huge spikes in demand during pandemic lockdowns. And the company had an added advantage: "contactless" delivery.

At its root, all of this is made possible by an autonomous driving system not unlike the one that keeps TuSimple's massive trucks from creaming other vehicles on the road, only vastly simplified. If TuSimple's trucks are the equivalent of a human—with our giant, sophisticated, calorie-devouring brains—then Starship's robots are more like a cockroach. Still a marvel of engineering and evolution, only of the economizing and reducing sort. Where a semitruck can generate enough spare electricity to power a whole rack of servers, Starship's mobile coolers must carry enough power in a single lithium-ion battery to drive themselves all day long, as well as sense, process, and respond to their environments.

A Starship robot does this with an array of nine cameras, giving it a 360-degree field of view. Six of these cameras are, like each of our eyes, part of a pair, giving the robot the ability to perceive depth. The three remaining cameras on the robot are "time-of-flight" cameras that can individually perceive depth in a very different way: by sending out a pulse of LED or laser light and measuring the time it takes for it to return to the camera. A laser-based time-of-flight camera is in fact a form of solid-state lidar, but the resolution of time-of-flight cameras has historically been much lower than what can be achieved with the lidar that goes on self-driving cars and trucks. The robot also uses ultrasonic sensors that send out a high-frequency pulse of sound and "listen" for its return. It's the same principle as echolocation in a bat or dolphin, only it's achieved in silicon.

What all these sensors have in common is that, thanks to their use in many common devices other than robots, they are widely available and relatively cheap. Cameras are by now ubiquitous, thanks in no small part to the supply chain that produces them for phones, and both time-of-flight sensors and ultrasonics are common in newer cars, where they are used in hazard, pedestrian, and back-up sensors.

These sensors also require very little power. The same is true of the "brain" of the Starship robot, which is a mobile application-optimized AMD Ryzen processor. Designed for laptops and other devices, it's significantly less powerful than a desktop computer, gaming laptop, or the kind of server-level equipment used on autonomous cars.

To cram object detection, navigation, SLAM, GPS, IMU navigation, and the like—all the things a self-driving car relies on, more or less—into such a tiny and energy-efficient package is by now a greater feat of engineering than autonomous navigation itself. Starship's robots are driving themselves with the computing power of an iPhone.

All of that economizing, all of that cramming of its systems into a package of technology that is good enough today and will only become cheaper with time, until the day it's almost disposably cheap, is what it takes to eke out a profit on two-dollar deliveries. At its most productive, each Starship robot is completing one delivery every thirty minutes and generating four dollars an hour in revenue. But I saw plenty of Starship robots idling on the campus of George Mason on the day I visited. Any time there are spikes in the demand for something, this kind of spare capacity can't be avoided. Whether it's ships or trucks or servers or delivery robots, idle machines are the price of satisfying peak demand.

Even so, every year, each of Starship's robots on the campus of George Mason drives a distance equivalent to a road trip from New York to Los Angeles. If we estimate that each robot is completing about one and a half deliveries per hour and working at most fourteen hours a day, then even if the robot can maintain that pace every day for an entire year, its annual "wage" is about $15,000.

That might seem like a lot for a device that doesn't pay taxes or rent, but that $15,000 must cover, among other things, maintenance. Those six solid rubber wheels, their bearings, and the motors and batteries that drive them wear out over the course of those 3,000 or so miles the robot covers in a year. Being electric, the cost of the energy to power the robot is negligible.

But then there's all the humans required to maintain and monitor each robot. Every place Starship delivers, it must hire a contingent of locals to remotely monitor its robots. This remote monitoring gets to both the key advantage and the greatest weakness of robot delivery systems: they don't actually have to be all that great at driving themselves autonomously.

All of the edge cases that keep the engineers at autonomous-driving companies awake at night don't really apply to a robot that weighs forty pounds and travels at a top speed of four miles an hour. In a collision with a vehicle, the Starship robot will happily crumple like the insulated plastic box it is. And as for encounters with pedestrians, well, the robot can always err on the safe side. Where an autonomous car must instantly decide if a piece of debris in its path is a life-threatening hazard or just a plastic bag blowing across the road, an autonomous delivery robot can stop and think a bit. And if it's still not sure, the delivery robot can call on one of its human minders to remotely connect through a wireless network, take a look at what it's seeing through its cameras, and make a decision for it.

Fundamentally, the fact that autonomous delivery is just so much lower stakes than autonomous driving is why this application of autonomous technology has arrived first. As a result, the problems Ryan says he is dealing with now are already outside the realm of technology and in the much more mundane and tractable arena of sales.

The maturity of the technology Starship is deploying is in part why Ryan took the job of head of business development—that is, deals—at the company. "So at the time, I thought aerial drones were way cooler," he says. "But you know, at the end of the day, you call me a sales guy, and I won't be offended. I need something that's ready, because it's really boring if I've got to sit around and wait for the engineers to keep working on it before I can go sell it." In January 2020, Starship reported its robots had completed one million autonomous deliveries.

That doesn't mean that Starship's only challenge is convincing college campuses, local restaurants, and city councils to accept its technology. The company also has to hire and train its robot minders, find people qualified to maintain their fleet, and—this has lately become one of the biggest bottlenecks—figure out how to make enough of its robots, fast enough.

Every hardware company must eventually confront the challenge of the vast gulf between making prototypes of its devices, usually built by hand, and making thousands or even millions of them. The former requires creativity, pluck, and a hacker mentality, while the latter requires, basically, Tim Cook. Apple's CEO rose to his position as Steve Jobs's anointed successor by being a master of the logistics of global manufacturing.

Starship, meanwhile, is still manufacturing its robots a few dozen at a time in the same place it always has, in Estonia, where the company's chief technology officer happens to live. For the company to scale up to the size envisioned by its founders, which could mean millions of delivery robots on the world's streets and sidewalks by the middle of the 2020s, it will have to manufacture them like automakers make cars and consumer electronics companies make phones—literally by the boatload, in the contract manufacturing capitals of East Asia and Southeast Asia.

If and when such robots become commonplace, they will be a remarkable form of infrastructure without precedent in our shared built environment. An autonomous delivery robot is, in essence, a conveyor that has escaped the factory or fulfillment center in which its progenitors evolved. It's a piece of the global, automated logistics network that extends, like a tendril miles long, from the rapidly multiplying array of warehouses in our midst to the doorsteps of our homes.

It's not hard to imagine that as those warehouses become more automated, a marriage of autonomous delivery and autonomous fulfillment will further drive down costs and increase the performance of Amazon and its kin. But it will probably take a very long time to make that transition: it turns out that, in terms of dollars spent and energy expended, a delivery truck piloted by a human is a remarkably efficient way to get things to their final destinations.

Chapter 22

THE LAST MILE

⟵ ───

In the time it takes me to get my seat belt unbuckled, Jenny Rosado, a UPS driver for thirty-one years, has turned off her truck, slipped the key from the ignition, set the parking brake, unbuckled her seat belt, risen from her seat, and pressed the button on a fob attached to her belt that remotely triggers the bulkhead door between the cab and the cargo area of the truck. By the time I have stood up from the passenger-side jump seat and am on my way down the two steep, diamond plate steel steps between me and the ground, she is at my back, package in hand, having already retrieved it from the back of the truck.

Her speed is no accident. UPS drivers are trained—drilled really, like soldiers repeating an action again and again, until everything is muscle memory—to go from stopping their truck to getting a package out of it in nine seconds.

"Today is an easy day," she says boisterously, already on her way to the customer's front door, me in tow. "Otherwise I'd be saying, 'Christopher, you need to move!'"

I'm in a picturesque suburb of Connecticut, where, per local regulations, the homes all sit on a minimum of one acre, and they go for around a million dollars apiece. It's a cool but not exactly crisp day in October, the sky overcast, rain sputtering on us now and again.

I'm wearing an official UPS uniform. I feel like a trainee on my first day. This is in part because of a phenomenon known as "enclothed cognition." Coined by a pair of psychologists in 2012, enclothed cognition describes how what we wear affects how we perceive ourselves, how we behave, how creative we are, how we process new information, and how we perform on tests of our skills. What I'm wearing also affects how others perceive me, of course. Throughout the day, customers who are home—and many of them are, since this is still pandemic time—assume I'm in training.

Jenny, who spends one morning each week teaching safety techniques to other UPS drivers, can't help but lecture me, in a good-natured way, about how to do the job. As the day wears on, I feel my journalistic detachment dissolving. Could I be a UPS driver? I wonder. Probably not, I think: This job is far more demanding than I realized. By the end of the day, my improper application of "the methods"—the seemingly endless list of optimizations, of thought and movement, practiced by experienced UPS drivers—has translated to a sharp pain in my left knee. It's an old injury that hasn't flared up in a decade, despite many miles of running. Delivering packages, it turns out, is even harder on my joints.

As the day wears on and pauses creep into the banter between Jenny and me, I have a chance to reflect. As I look at the interior of the UPS truck, which is all flat planes, hard angles, and stainless steel, like the faceted superstructure of a World War II–era battleship, it strikes me that there is something steampunk about the way that packages are delivered on the last mile of their journey. Despite the veneer of slick, internet-era modernity companies like Amazon have laid atop modern commerce, with one-click shopping, ever-faster delivery, and seemingly infinite selection, there is a hard, physically demanding reality underneath.

Moving all the stuff we buy online along the final leg of its journey is a Willy Wonka–ish hybrid of two things, one constantly rushing into the future and the other inseparable from the past. On the one hand, all such deliveries are directed from afar by a mysterious, invisible, and all-seeing artificial intelligence, constantly revising answers to mathematical problems so hard that their definitive solutions cannot be found by any computer in existence, even if it should continue running for billions of years. And then there's the moment-to-moment, almost balletic, physical and mental calculus required of a single human being.

The people who take our packages on the final legs of their journey are, as physicians, physical therapists, and sports medicine specialists have dubbed them, "industrial athletes." There is no better way to describe human beings who are asked to make an average of 135 stops a day, and a maximum of 160, representing upwards of 200 packages, in a single ten- or even twelve-hour shift with only one break for lunch, for weeks, months, even years on end, without destroying their body.

Jenny, fifty-three and sturdy, with dark salt-and-pepper hair buzzed on the sides, moves with the kind of relentless confidence that comes from a reserve of endurance and functional strength built up over decades. She is the only woman driver at her delivery station in Stratford, Connecticut, a facility built in 1966 and in more or less continuous operation ever since. Its walls, floors, and fittings are blackened by decades of use, like the inside of some Victorian factory.

There have been other women drivers in Stratford, she allows, but after a couple of years they tend to get injured and can no longer do the job. It's sad, she continues: usually it's a person's knees that go. After a single day on a route with "only" ninety stops—so we'll have time to talk, says Jenny—I can see how that could happen even to someone young and fit.

As we cruise down a winding suburban road, the sidewalks slick with leaves, fall colors blazing, Jenny distills the many lessons she and others at UPS teach new drivers in order to make them as productive and, she emphasizes, as safe as possible. "You never want to work harder than you should," she says. "You see a lot of people come out here and they work *way* too hard. They don't follow the methods."

UPS is all about its methods. If you think your job is regimented, I invite you to read the seventy-nine-page "Service Provider Delivery and Pickup Methods" section of the UPS *Industrial Engineering Standard Practice Manual,* an internal document the company regularly updates that describes the 340 different methods a UPS driver is supposed to internalize in order to be successful. They range from how to load your next five delivery stops into your working memory so you can preplan how you'll accomplish them even as you're walking to the front door of your current drop-off to how to enter and exit a truck in a way that minimizes the strain on your body.

Inexperienced drivers, and those who are hired on a provisional basis who may never make the cut, struggle with the mental skill of organization, Jenny continues. Both Jenny and the manual emphasize that you should never touch a package in the back of a truck more than once before grabbing it to make a delivery.

In theory, all packages in the back of the truck have already been organized by dedicated preloaders, whose shifts start at 3:00 a.m. They are the invisible partners of the driver and either beloved or reviled by them, depending on how good they are at their jobs. In reality, stuffing enough packages of various sizes, many too large to be put on a shelf, into the back of a UPS truck is a bit like loading a dishwasher after Thanksgiving dinner. Considerable creativity is required just to get all the boxes to fit. Organization can fall by the wayside, and small items must necessarily be secreted around and behind larger ones.

So it falls to the driver to periodically reorganize the back of their truck to assure that all packages, even big ones that must lie on the floor, are in sequential order, labels facing out. The goal is that on most stops a driver should be able to grab the next package they'll need without venturing more than thirty inches—precisely thirty inches—into the back of the truck.

Everything about a UPS delivery driver's job is about finesse, about metronomic precision. To paraphrase Flaubert, it's about being regular and orderly in as many tasks as possible so you can improvise when appropriate. Delivering packages is a game of inches—or really, seconds, since shaving the slightest amount of time off of anything you do hundreds of times a day means the difference between getting all your packages delivered and the bosses having to send someone out to rescue you by taking a few dozen off your hands.

Inexperienced drivers, trying not to fall behind, will hurry, says

Jenny. But in this job haste is the enemy of focus, and focus is key. Here's section 6, point II, clause C of the "Service Provider Delivery and Pickup Methods": "Do not run. Running requires too much of your attention. Remember that your primary objective is to give your attention to planning ahead."

"They're fighting the load; they're fighting the job," says Jenny. "And others try to make it up by speeding." She shakes her head. Speeding while delivering packages invites disaster. For delivery drivers especially—some of whom cover tens of thousands of miles a year, for years, in oversize vehicles with poor maneuverability and worse visibility, at times of day when traffic can be particularly heavy, and in locales that aren't exactly friendly to trucks—speeding, or any other unsafe driving practice, can make an accident inevitable. And when dealing with something the dimensions of a truck, those accidents can be worse than costly—they can be lethal.

Take, for example, the case of Joy Covey, a single mother of a young child and a board member of the National Resources Defense Council. In 2013, she found herself cycling not far from her home, down Skyline Boulevard, just outside Palo Alto, one of the hubs of Silicon Valley. Suddenly, a white minivan heading in the opposite direction made a sharp left turn in front of her. It was too late for her to stop. Her body slammed into the side of the van.

But who was Joy Covey, other than a recently retired technology executive who had apparently been looking forward to living off the proceeds of her years of hard work and raising her young son?

In a tragic irony of Shakespearean proportions, Covey was the former chief financial officer of Amazon. She had been recruited by Jeff Bezos himself. During one of the most critical periods of the company's early growth, she had been his right-hand woman.

The van that cut her off was owned by a subcontractor of Amazon. The driver had been delivering packages. Covey died at the scene.

Jeff Bezos spoke at her funeral, according to the book *The Everything Store* by Brad Stone. Choking up, he said: "Joy and I talked often about a day in the future when we would sit down together with our grandkids and tell the Amazon story."

A 2019 investigation by ProPublica and *BuzzFeed News* documented a years-long pattern of Amazon's subcontracted last-mile delivery companies pushing drivers to accomplish unreasonable delivery goals on timetables that forced many to cut corners, ranging from peeing in bottles because they didn't have the time to take proper bathroom breaks to reckless driving. Amazon pushed back, calling the report "another attempt . . . to push a preconceived narrative that is simply untrue. Nothing is more important to us than safety." A representative for the company continued: "Unfortunately, statistically at this scale, traffic incidents have occurred and will occur again, but these are exceptions, and we will not be satisfied until we achieve zero incidents across our delivery operations."

Subsequent reporting by a variety of news outlets, including Vice, the *New York Times*, and NBC News, found that whatever Amazon's protestations about its commitment to safety, the fundamental business logic pushing drivers up to and beyond their limits was in full effect both before and especially during the strained times of the pandemic. As of late 2020, drivers reported they still felt enormous pressure to finish all their deliveries before the end of their shifts, whatever it took.

▶ ▶ ▶

UPS has more than 123,000 vehicles worldwide, and it is inevitable that, every year, some are involved in fatal accidents. But the tension between productivity and safety in the company is resolved within a system of incentives that is markedly different from Amazon's.

Jenny, for one, is no stranger to huge workloads, especially during peak times. "Peak" used to describe the holiday rush of packages, but from the moment pandemic lockdowns began in the United States in March through the 2020 holiday season, which one FedEx executive described as "peak on top of peak," this period of furious activity became the new normal for the entire logistics industry.

Despite the increased workloads, Jenny tells drivers there's no reason to stress: they're getting paid by the hour, after all. For any overtime beyond their usual eight-hour workday (plus an hour for lunch), they are paid, per regulations, 50 percent more than their usual wages. UPS drivers grumble about the strain of peak, but they also will tell anyone who asks that the best thing about it is the money.

By industry standards, UPS drivers are paid very well. Because of her decades of experience, Jenny has reached the highest pay tier a driver can reach: she makes $38 an hour. She pays nothing for her health insurance. She gets a pension and six weeks a year of vacation. Amazon drivers, meanwhile, are for the most part contractors who start at $15 an hour, below even the $20 hourly wages of temporary, seasonal UPS drivers.

Aside from a few highly skilled, highly specialized forms of delivery, no other drivers in the last-mile industry are compensated the way UPS drivers are. The most obvious reason is that nearly all permanent UPS drivers belong to a union, and not just any union, but the Teamsters. This is, after all, the organization whose mid-twentieth-century heyday was depicted in the movie *The Irishman*, starring Al Pacino as Teamsters boss Jimmy Hoffa. In real life, Hoffa was infamous for allying with organized crime and was later killed by a Teamster who was also a hitman.

The Teamsters are but a shadow of their former power, but they are still led by a Hoffa—James P. Hoffa, son of Jimmy—and are an organization not to be trifled with. In 1997, 180,000 unionized UPS workers went on strike for sixteen days, shutting down the company and extracting major concessions from management, both at the time and for decades to come. That was a crazy time, and a hard one, says Jenny. In order to make up for the shortfall in her wages, she took a temporary job helping a landscaper. It was the only thing she's ever done that was harder than being a UPS driver.

Jenny's local union is number 191, and these days, she says, "they get along with the company. They're able to talk to each other. Where I know there are a lot of other unions that don't have that kind of relationship."

UPS's relationship with its drivers' union is ancient, by the standards of modern corporations, and goes back to the 1930s. The

Teamsters have often been accused of "business unionism," that almost uniquely American tradition of treating unions not as a means of worker liberation but as businesses in their own right, tendentious but not ultimately opposed to the profit-making enterprise at the heart of American capitalism.

By the standards of, say, Bolshevism, the Teamsters are, politically, positively hard right. But just by continuing to exist, the Teamsters have become something unusual. We now live in an era in which Mao's Cultural Revolution has been transformed into state-sponsored hypercapitalism, and many European countries treat the design of competitive markets as a high art, on a par with their production of luxury goods. In the United States, meanwhile, membership in labor unions has dropped to around 11 percent of all workers, a level not seen since the Great Depression.

In last-mile delivery, none of UPS's primary competitors have unionized workforces. (DHL is the exception, but it claims only a small fraction of the U.S. market for shipping.) FedEx subcontracts its delivery areas to individual companies that operate like fast-food franchisees.

Amazon has adopted the same model as FedEx. Everything about Amazon's decision to hire delivery companies that hire drivers, rather than hiring those drivers directly, is about pushing down wages, eliminating workplace protections, evading liability in the event of accidents, avoiding workplace litigation, eliminating the expense of benefits, and eliminating the possibility of drivers ever unionizing, says David Weil, a professor at Brandeis University.

David coined the term "fissured workplace" in order to describe a trend he observed, stretching back decades, in which businesses hire subcontractors to perform even business-critical tasks that they once hired employees to do. The classic example is the way countless companies stopped hiring janitors and hired companies that hired janitors, severing the legal relationship between the company and workers that were once its employees. This has led to a broad spectrum of abuses, David alleges, from wage theft and sexual harassment to unsafe working conditions.

Amazon sets safety standards and expects its subcontracted delivery companies to abide by them, using the threat of "firing" those companies if they fail to enforce those standards. The company periodically dissolves its relationship with subcontracted delivery companies that cut corners, don't meet its high standards for on-time delivery of packages, or violate its safety standards. One such purge in February 2020 resulted in layoffs for 1,300 drivers across the United States, out of a total population of 75,000 drivers working for more than 800 different companies.

But legally, Amazon has no obligation to provide for the health and safety of drivers, says David. It's even worse for Amazon's "Flex" drivers, who deliver packages in their own personal vehicles and sign up for deliveries in much the same way an Uber or Uber Eats driver picks up fares and deliveries. These drivers "are not covered by OSHA, they are not covered by worker compensation laws or by unemployment insurance," he adds. "If you are one of those drivers and someone at Amazon sexually harasses you, you have no recourse."

For three years, under President Barack Obama, David was head of the Wage and Hour Division of the U.S. Department of Labor, which enforces the Fair Labor Standards Act. "One of the great battles we became involved in was exactly this model," he says. "It is incredibly corrosive to our entire system of protecting the workers at the bottom who are most exposed to abuse. The reason we have labor laws in the first place is that they are the group of people least able to protect themselves."

In cases brought against Amazon, the company has denied any responsibility for accidents caused by its drivers, on the grounds that those drivers are employees of its subcontracted delivery companies, not Amazon itself. Amazon's arguments in court make plain how it has crafted contracts with its delivery "partners" that preclude the possibility that Amazon would ever be classified under the law as a "joint employer" of those drivers. Thus, incredibly, workers wearing uniforms bearing the Amazon logo, in vans with six-foot-wide smiles on their side, dispatched from Amazon delivery stations, carrying mobile devices running Amazon's custom mapping and routing software, and delivering nothing but Amazon packages are not legally Amazon's employees, under current law and given current legal precedents.

Guidance issued by the Department of Labor, under the Trump administration, further eroded even the last flimsy protections afforded such workers by eliminating the traditional sternly worded memos the department once issued when accusing a company of violating labor laws. Now companies, such as Amazon, that leverage the fissured workplace to eliminate responsibility for their workers' safety and actions do not face even the threat of public shaming, aside from the odd brickbat lobbed by the handful of investigative journalists who follow the company closely.

It's not hard to see how Amazon's—and for that matter, FedEx's—fissured workplaces protect these companies from liability. But it can be harder, absent an understanding of the shrinking number of counterexamples like UPS, to understand how it also insulates them from any financial or other sort of responsibility for those employees.

Here's how David puts it: UPS is a for-profit company, "but they have a direct incentive to reduce injuries, because those injuries are on them." Being responsible for drivers' health-care costs, they must work on keeping drivers safe because they end up paying for treatment of those injuries. In addition, while companies that rely on subcontractors can count on high turnover to push injured workers out of their pool of employees, UPS invests more in each driver.

In part, this is a necessity: UPS has spent hundreds of millions on perfecting its technology, but in some ways it is saddled with outmoded legacy systems. It must compete on other dimensions of efficiency, such as highly trained drivers who know their routes well. This is a further disincentive to lose those drivers and a further incentive to keep them healthy. The company has perfected time and motion studies not only for how every task should be performed—those 340 methods—but also for each individual route a driver completes.

"You had this environment because UPS owns this health and safety system, and the responsibilities for drivers, so they optimize the system around that," says David. In a franchise model such as

Amazon's, by contrast, "the only potential workers' compensation claim that could be brought is on this low-level franchisee," he adds.

Where great American retail companies of previous eras, such as Sears, Roebuck and Co., were built on a principle of mutual responsibility and shared ownership between employees and the firm, Amazon has in many ways chosen a different route. It's one embodied by the management philosophies of one of Jeff Bezos's self-confessed idols, Sam Walton, founder of Walmart. America still has big companies in which workers have leverage and in which companies use some of their windfall profits to take care of their employees, but increasingly, this is a privilege reserved only for white-collar workers.

▶ ▶ ▶

UPS has calculated that every mile it saves per driver, per year represents $30 million in avoided expenses. Between 2013 and 2019, the company went from delivering 4.3 billion packages worldwide to 5.5 billion. The bigger UPS gets, the more a single math problem, one with the potential to shave miles and minutes off of all those routes, matters to the company.

The traveling salesperson problem, first formally described by mathematicians in the 1930s, goes something like this: Imagine a traveling salesperson needs to go to five different cities. What's the shortest route between those cities that gets them to all five and back to their starting point? Visualizing this problem, it can seem trivial to solve an exercise in connecting the dots that even a child could find the best possible solution to. But assuming you got the right answer, in so doing, you would have unknowingly identified the best solution out of 120 possibilities.

Add a sixth city, and the shortest route is but one of 720 different possibilities. Again, this isn't a hard problem to solve if the cities are all strung out more or less in a line, but the moment they become a random cluster of points, a spatter of dots on a map, the solution becomes much more challenging to identify.

Now consider the day of an average UPS driver: How best can we optimize their route? If six destinations yields 720 possible routes and only one is the best, how many does 135 destinations translate to? Approximately this many: 269,000,000,000,000,000,000,000,000, 000,000,000,000,000,000,000,000,000,000,000,000,000,000,000, 000,000,000,000,000,000,000,000,000,000,000,000,000,000,000, 000,000,000,000,000,000,000,000,000,000,000,000,000,000,000, 000,000,000,000,000,000,000,000,000,000,000,000,000,000,000, 000,000,000,000,000,000,000,000,000,000,000,000,000,000,000, 000,000,000,000,000,000,000,000.

To put that number in context: if every single atom in the universe were itself a little universe with an equal number of atoms inside, and you counted up all the atoms in your hypothetical universe of universes, this number is still bigger. To even approach the size of this number, you would need to multiply your previous result by another preposterously large number, one greater than the total number of atoms in planet Earth.

OK, you might be saying, maybe there are rules that allow us to reduce the size of the calculation required to find the best route among 2.69 times 10^{230} possibilities. Now you have stumbled on one of the most fundamental questions in computer science, one that has vexed mathematicians and engineers alike since it was first stated in 1971, and we still haven't delivered a single package. So far as we know, there is no way to find the absolute best solution to the traveling salesperson problem without trying every single possible route.

This problem is so challenging, and in its broader formulation so fundamental to so many other areas of computation and mathematics— from DNA sequencing to the design and manufacture of microchips— that if you can prove there's a more efficient way to solve it, you'll win $1 million from the Clay Mathematics Institute in Peterborough, New Hampshire.

Now let's make the problem reflect the real world, which means making it even more complicated. To solve this problem for our UPS driver, or for Jenny, specifically, on the day she's delivering our USB charger to a homebound office worker, other factors must be added to the equation. To name but a few, they might include stops that must be made before 10:30 a.m., the cutoff for delivery of next-day packages; portions of the route where a driver would prefer not to turn left on account of safety; the need to make today's route resemble yesterday's, since UPS has found this is critical to keeping drivers happy and efficient; and the availability of other drivers and trucks to complete some portion of the stops on this route. Every day, UPS must solve not merely Jenny's traveling salesperson problem but many others, all of which also exist within a greater matrix of other, network-wide considerations. Now consider that traffic changes throughout the day, so you'll probably want to recalculate this route fairly often, perhaps even once every few minutes, since the time, if not the distance, between stops will grow longer if there's congestion.

The solution to this literally insoluble problem is UPS's On-Road Integrated Optimization and Navigation (ORION) system, an algorithm of algorithms that attempts to take all this into account when plotting a route for a driver. Released in 2013, ORION didn't get dynamic updating of routes—the function most of us take for granted on our Waze or Google or Apple Maps apps that recalculates our best route based on traffic—until 2020. ORION also didn't start delivering to drivers that staple of all driver GPS systems, turn-by-turn directions, until 2018.

The company points out that both of these functions came later to their systems than to consumer mapping apps, because of the complexity of the problem of directing a driver on a route with hundreds of stops, many of which are in unusual and very specific locations, such as the loading dock at the back of a business in a strip mall.

It's been estimated that UPS spent between $200 million and $300 million to develop ORION, and an unknown amount of money since then in order to continue refining it and adding new features. Jenny uses it, but for all the company has spent on it, she also has a habit of ignoring it, especially when it asks her to make left turns.

One piece of UPS lore that refuses to die is that UPS drivers don't make left turns because waiting to yield to oncoming traffic is inefficient. It turns out that what was once a useful rule of thumb has yielded to the algorithmic overmind, which has decided that, sometimes, left turns are desirable.

Jenny disagrees. "Most of the time we won't do it if we don't think it's safe, especially if we know we're coming back." The goal is to hit at least 80 percent of your stops in the order that the system dictates, which means that on some days drivers like Jenny are, despite all that investment, still disobeying ORION on one out of every five deliveries.

ORION, it turns out, is a good example of where artificial intelligence and human intelligence can and do mesh. The human has to deal with their immediate reality, everything from how packages have been stacked in the back of their truck to what traffic is like at the precise moment they must decide whether to make a turn. And the system is not all-knowing. Surprisingly, UPS's routing algorithms don't take into account how big a package is, says Jenny. If she has an eight-foot-long, hundred-pound boxed trampoline in the back of her truck—trampolines were popular along her route at one point during the pandemic, when the weather was warm and parents were desperate to keep stay-at-home kids occupied—she's going to get rid of it as early as she can in the day so she's not tripping over it on the rest of her deliveries.

Plus, a large body of research shows that humans are remarkably good at solving the traveling salesman problem on their own. Scientists aren't sure why, but something about the way we process visual information makes us pretty decent at finding a nearly optimal route through even a dense thicket of potential stops. Add to that a driver's local knowledge, and it makes sense that the best possible way to accomplish a given delivery route is to give experienced drivers some amount of discretion in how they accomplish their drop-offs.

The way ORION and the rest of UPS's systems communicate with drivers is through each driver's DIAD, which stands for delivery information acquisition device. The DIAD is like a smartphone from an alternate dimension in which Steve Jobs never killed off the Black-Berry by offering consumers a slim, button-free slab of glass known as the iPhone.

The current generation DIAD, the DIAD V, weighs 1.3 pounds. If you stacked four iPhones on top of one another, you'd get something approximately the weight and dimensions of this DIAD. Instead of a touch screen, it has a gigantic array of buttons below its tiny, first-generation smartphone-size screen. First introduced in 2013, it's a device better known for durability than speed.

Which is fine, for the most part. Updates to the DIAD have mostly come in the form of more power and features delivered from the cloud—that ORION system and its turn-by-turn directions, for example. Most of the time, the DIAD just needs to display a list of packages and their addresses, along with a cryptic series of codes denoting the nature of a pickup or drop-off and a field for notes. That's where drivers warn each other about hazards at particular stops—mostly dogs.

Dogs are not a problem for Jenny, but they might be a problem for anyone who might have to take over her route, because the dogs she encounters almost every day know her well. She gives them treats, which she's not supposed to do. Management doesn't want drivers doing anything that might attract a dog's attention, even the good kind.

At one stop, after we drop off a package and chat with a customer, a boisterous Labrador retriever bursts out the front door of the house and jumps onto the truck. Jenny tells me to lift my legs so as not to get in the dog's way—this is a routine she and the Lab have developed. To get it out of the truck, she whistles, tosses a treat twenty feet down the driveway, and then throws the truck into gear and speeds away before the Lab can heave its giant, enthusiastic body back onto the truck.

The original version of the DIAD was so big, says Jenny, that "it was like a weapon! You should have seen that big-ass calculator we had." The first-generation DIAD rolled out in 1990 and, indeed, it was truly gigantic. There is a press photo of one person passing it to another, and it unintentionally makes it look as if it takes two sets of hands just to hold it up. As big as a breadbox and top-heavy, it was, for its time, a technological marvel.

The DIAD III had an audio modem on it. Drivers would sometimes ask to borrow a customer's phone, then call a special number and shove the phone's receiver against the face of the DIAD, which would emit a series of beeps, squawks, and hisses in order to transmit data back to the company's central mainframes. It came out in 1999, two years before Intel would release its first CPU intended specifically for laptop computers.

Codeveloped with Motorola, the DIAD III also had something truly remarkable for its time—a cellular modem that worked everywhere. To accomplish this, UPS had to get cellular carriers all over the country to agree to collectively provide coverage everywhere they had towers. It was the year the BlackBerry was first released; the iPhone was still eight years in the future. In other words, UPS tied with BlackBerry's maker RIM, and a handful of cell phone makers in Japan, in the race to release the world's first commercial-scale, cellular-enabled personal communications device.

While the DIAD may be the sole way UPS sends information to drivers when they're out on their routes, it's far from the only way drivers are communicating with UPS, and most of those methods of communication are indirect. A modern UPS truck is bristling with sensors. UPS's computers, and every driver's supervisor, knows precisely where a driver has been, how quickly they drove, if their bulkhead door was open or their seat belt unbuckled when they were in motion, and if they began any part of their journey by putting their truck in reverse—all no-nos.

When UPS first began gathering this kind of data, drivers feared it would be used as yet another means of control. Some saw it as another way to disempower them and, ultimately, to make their job less complicated and individual drivers more interchangeable and disposable. All of that, aside from the disposability of drivers, has come to pass. But this level of driver surveillance has also meant more compliance with safety regulations the company says are critical to reducing accidents and injuries.

Before I know it, we're at our ninetieth stop of the day. I haven't looked at my watch once, except when we took a break for lunch. It's not as if the day went by quickly—often, I looked into the back of the truck and despaired that the mountain of packages was hardly diminished by all our efforts. Yet somehow, all the activity also eradicated my sense of time passing. This is typical, says Jenny, especially when you get in the zone.

I don't know what was in the last package of the day. But it was a small one, a rectangular box with dimensions closer to a cube than a sheaf of paper. Maybe it contained a USB charger. I know we'd already delivered electronics earlier in the day—one woman was delighted to receive her brand-new iPhone on the same day that model was released.

So let's say it was a USB charger, after all. No one was home. We walked to the front door, doing everything by the book—scanning our walk path for obstructions, walking briskly. ("A brisk pace commands attention," notes the manual, in a tiny haiku that sums up every impression I've ever gotten from a UPS driver, but especially as they strode up to my desk or my front door, martial in their uniform and bearing.)

We left the package tucked up against the side of the house, in case of rain, but not directly in front of the door, because you never want the customer to trip over it. We walked away.

In the hypothetical journey of our USB charger, it's March 23. While the person who needs it to keep their kid's device going ordered it only yesterday, this is the end point of a much older, much longer journey. It's been almost exactly two months since this charger left a factory in Vietnam. It has traveled more than 14,000 miles, across twelve time zones, by truck, barge, crane, container ship, crane, and truck again, all before it trundled down a few hundred yards of conveyor, flitted about on the back of a robot, and was ferried again on, all told, miles more conveyor and at least two more trucks, before being hand-carried to someone's front door.

Along the way, the charger was touched by dozens of people, all of whom had their humanity, their hopes and fears, their dreams and histories, trained and channeled into achieving a maximum of efficiency, a machine-like efficiency, in order to keep pace with the automation they work alongside. It's March 23, 2020, barely a couple of weeks into the new normal for the global logistics system, the never-ending "peak" that would strain every link in the chain between where things come from and where they go.

And tomorrow, and the day after, and the day after that, every person and machine in that system will do it again. Months or years later, as you read this, they are still doing it, for you and I and billions of other people on the planet. As a drop of rain evaporates from the ocean and one day returns to it, you've witnessed the path of a single object through this system. Now multiply that by 100 billion—the number of parcels shipped every year, worldwide, as of 2020. And then imagine a future in which that number has doubled or tripled; imagine a future in which it is the way virtually every finished object gets anywhere.

ACKNOWLEDGMENTS

To admit a journalist into your company, facility, or inner world is to take a leap of faith for which I'm not sure there's any ready parallel in other areas of life. The people who deserve the most thanks for the existence of this book, but who are in no way responsible for any errors it contains, are every named source in the book, and many others whose names did not make it in, either because we spoke on background or because there is only so much that can be crammed into one book. Not only were dozens and dozens of interviewees generous with their time, but many extended that generosity by reviewing explanatory portions of the book that were based on our conversations as well as materials they provided. Thanks to their frank feedback, many of my embarrassing oversights, clunky descriptions, and not-quite-right analogies were rendered more faithful to their respective realities. Thanks are also due to my editors at the *Wall Street Journal*, who both granted me time off to complete this project and, from its inception, expressed their enthusiasm for it. Those same editors have, over the past seven years, given me the most phenomenal education in how to be a journalist that I could have asked for. Through their unwavering support, they have helped me develop in ways I didn't know were possible, while also never hesitating to give me a kick in the pants when I needed it. There are too many to name, but if you are reading this: You know who you are.

This book would not exist without the support of close friends, whose endless chains of texts with me about my travels, struggles, and ideas are all the receipts they need to prove that I owe them big-time.

From the start, Rafe Sagalyn, agent extraordinaire, has been a thought partner of that rarest sort—one capable of both seeing the bigger picture and motivating the fragile writerly ego by making it think the idea belonged to it all along. By the same token, my editor, Hollis Heimbouch, deserves credit for being a clear and deep thinker, skilled at her trade, who has had a profound impact on this text—not to mention a much-needed well of encouragement and enthusiasm throughout the writing process.

Every editor and educator who ever demanded more than a few hundred words from me deserves some credit for this book, from my high school English teacher Ms. Peacock, wherever you are now, to Professor Melvin Konner, whose undergraduate writing seminar in psychology and anthropology inspires me to this day.

Finally, this book would have been a complete non-starter if not for the support and labors of Shep, every pint-size member of our immediate family, and every member of our extended pandemic pod. They say it takes a village to raise a child; it turns out the same is true for creating a book.

NOTES

Chapter 1: The Gathering Storm

10   17,600 different products: Jon Delano, "Shortage in Chinese Products Could Mean Empty Shelves in U.S," CBS Pittsburgh, February 28, 2020, https://pittsburgh.cbslocal.com/2020/02/28/chinese-products-shortage-coronavirus.

10   artificial sweeteners: Dion Rabouin, Joann Muller, Bob Herman, and Courtenay Brown, "Brace for Coronavirus Supply Shocks," Axios, March 12, 2020, https://www.axios.com/brace-for-coronavirus-supply-shocks-d9bef456-e72f-4cdd-aa1d-4e55b87c246c.html.

10   three in four U.S. firms: Dion Rabouin, "Coronavirus Has Disrupted Supply Chains for Nearly 75% of U.S. Companies," Axios, March 11, 2020, https://www.axios.com/coronavirus-supply-chains-china-46d82a0f-9f52-4229-840a-936822ddef41.html.

12   cod caught off Scotland: "Scotland to China and Back Again . . . Cod's 10,000-Mile Trip to Your Table," *The Herald* (Scotland), August 21, 2009, https://www.heraldscotland.com/news/12765981.scotland-to-china-and-back-again-cods-10000-mile-trip-to-your-table.

12   quartz harvested in Appalachia: Vince Beiser, "The Ultra-Pure, Super-Secret Sand That Makes Your Phone Possible," *Wired*, August 7, 2018, https://www.wired.com/story/book-excerpt-science-of-ultra-pure-silicon.

Chapter 2: The Box

15   51 percent: Nicholas Prescott and Jennie Litvack. "Viet Nam—Poverty Assessment and Strategy," World Bank, *Understanding Poverty*, July 1, 2010, https://documents.worldbank.org/en/publication/documents-reports/documentdetail/904581468761686067/viet-nam-poverty-assessment-and-strategy.

15   10 percent: Obert Pimhidzai, "Climbing the Ladder: Poverty Reduction and Shared Prosperity in Vietnam." World Bank, *Understanding Poverty*, April 4, 2018, https://documents.worldbank.org/en/publication/documents-reports /documentdetail/206981522843253122/climbing-the-ladder-poverty -reduction-and-shared-prosperity-in-vietnam.

16   Samsung smartphones: Adnan Farooqui., "Where Are Samsung Phones Made? It's Not Where You Think," SamMobile, November 26, 2019, https://www .sammobile.com/where-are-samsung-phones-made.

16   Google also shifted manufacture: Cheng Ting-Fang and Lauly Li, "Google to Shift Pixel Smartphone Production from China to Vietnam," Nikkei Asia, August 28, 2019, https://asia.nikkei.com/Spotlight/Tech-scroll-Asia/Google -to-shift-Pixel-smartphone-production-from-China-to-Vietnam.

20   Nearly 800 million units of container shipping: "Container Port Traffic (TEU: 20 Foot Equivalent Units)," Data, World Bank, 2020, https://data .worldbank.org/indicator/IS.SHP.GOOD.TU.

22   quarter of the world's shipping volume: Patrik Berglund, "Ocean Shipping Alliances: Do They Still Matter?" Xeneta, April 22, 2016, https://www.xeneta .com/blog/shipping-alliances-do-they-still-matter.

23   3.7 million units: Sam Whelan, "Vietnam's Cai Mep Port Gets Set to Benefit from Supply Chain Switch from China," *The Loadstar*, July 1, 2020, https:// theloadstar.com/vietnams-cai-mep-port-gets-set-to-benefit-from-supply -chain-switch-from-china.

23   about a third of its capacity: Sam Whelan, "Bigger Vessels Will Secure a Brighter Future for Vietnam's Cai Mep Container Terminals," *The Loadstar*, May 8, 2015, https://theloadstar.com/bigger-vessels-will-secure-a-brighter -future-for-vietnams-cai-mep-container-terminals.

Chapter 3: Ships and Other Cyborgs

25   travel by ship: Rose George, *Ninety Percent of Everything: Inside Shipping, the Invisible Industry That Puts Clothes on Your Back, Gas in Your Car, and Food on Your Plate* (New York: Picador, 2014).

25   according to the United Nations Conference on Trade and Development: "World Seaborne Trade," United Nations Conference on Trade and Development (UNCTAD), December 7, 2020, https://stats.unctad.org/handbook /MaritimeTransport/WorldSeaborneTrade.html.

26   70 percent of the value of all goods: UNCTAD, "World Seaborne Trade."

26   Gyeongsangnam-do, South Korea: "Vessels: OOCL *Brussels*," OOCL, https:// www.oocl.com/eng/ourservices/vessels/mclass13208/Pages/ooclbrussels .aspx.

26   Hyundai Merchant Marine *Algeciras*: "HMM Names World's Largest Container Vessel, 'HMM *Algeciras*,'" Marine Insight, https://www.marineinsight .com/shipping-news/hmm-names-worlds-largest-container-vessel-24000 -teu-giant-hmm-algeciras-at-dsme-shipyard/.

27 trapped at sea: "400,000 Seafarers Stuck at Sea as Crew Change Crisis Deepens," International Maritime Organization, https://www.imo.org/en /MediaCentre/PressBriefings/Pages/32-crew-change-UNGA.aspx.

36 efficiency and reliability above all else: "World's Largest Engine Powers the World's Largest Container Ship," Marine Insight, January 16, 2017, https:// www.marineinsight.com/shipping-news/worlds-largest-engine-powers-the -worlds-largest-container-ship.

38 conscripted into forced labor: Ian Urbina, *The Outlaw Ocean: Journeys Across the Last Untamed Frontier* (New York: Alfred A. Knopf, 2019).

39 recalled in a video: Jeff Tsang, "WE NEARLY CRASHED! OUR ENGINE BROKE!? A Close Call with Pack of 200+ Ships Outside Shanghai," YouTube, March 22, 2020, https://www.youtube.com/watch?v=h1-wbV8PkmI.

41 superaccurate marine clocks: Dava Sobel, *Longitude: The True Story of a Lone Genius Who Solved the Greatest Scientific Problem of His Time* (London: Fourth Estate, 2014).

41 age of the *Titanic*: "History of Sperry Marine," Sperry Marine, https:// www.sperrymarine.com/corporate-history/sperry-marine.

42 electromechanical device: "'Metal Mike' Guides the Queen Elizabeth," *New York Times*, September 26, 1946.

Chapter 4: Coming to America

46 around twenty-four hours: "Average U.S. Container Vessel Dwell Times, 2016," Bureau of Transportation Statistics, https://www.bts.gov/content/average -us-container-vessel-dwell-times.

46 Built in 1917: "Port of Los Angeles Offers Historic 'Warehouse No. 1' on the LA Waterfront for Commercial Development Opportunity," Port of Los Angeles, March 3, 2020, https://www.portoflosangeles.org/references/news _030320_warehouse_one_development.

47 $434,000 a year: Jack Dolan and Paul Pringle, "How One of L.A.'s Highest-Paying Jobs Went to the Boss' Son," *Los Angeles Times*, June 11, 2016, https:// www.latimes.com/local/lanow/la-me-adv-port-pilots-snap-story.html.

47 nearly as much: Gloria Hillard, "Harbor Pilots Reap High Rewards for Dangerous Job," NPR, March 21, 2012, https://www.npr.org/2012/03/21 /149091141/harbor-pilots-reap-high-rewards-for-dangerous-job.

48 physically and historically: Nathan Masters, "Why Is SoCal's Harbor Split Between Two Cities?" KCET, November 21, 2017, https://www.kcet.org /shows/lost-la/why-is-socals-harbor-split-between-two-cities.

48 the largest port: Eric Johnson, "JOC Rankings: US–China Trade War Accelerates Market Share Losses for West Coast Ports," JOC.com, May 4, 2020, https://www.joc.com/port-news/us-ports/joc-rankings-us%E2 %80%93china-trade-war-accelerates-market-share-losses-west-coast -ports_20200504.html.

48 half of all the goods shipped to America from Asia: Bill Mongelluzzo,

"Top US Port Gateways Expand Share of Asian Imports," JOC.com, August 31, 2018, https://www.joc.com/port-news/us-ports/port-los-angeles/top-us-port-gateways-expand-share-asian-imports_20180831.html.

48   507 container ships: "Operating Performance Improved Significantly with Clear and Firm Development Strategy; Cosco Shipping Holdings Announces 2019 Annual Results," Cosco Shipping, March 31, 2020, https://www.accesswire.com/583285/Operating-Performance-Improved-Significantly-with-Clear-and-Firm-Development-Strategy-COSCO-SHIPPING-Holdings-Announces-2019-Annual-Results.

48   100 million tons of deadweight: "Company Profile," Cosco Shipping, February 28, 2019, https://world.lines.coscoshipping.com/newzealand/en/aboutus/companyprofile/1/6.

54   one in twenty chance of dying on the job: Susan Salisbury, "The Lucrative, Dangerous Job of Being a Harbor Pilot," *Palm Beach Post*, September 24, 2016, https://www.palmbeachpost.com/article/20150422/BUSINESS/812063254.

Chapter 5: Parallel Parking 1,200 Feet of Ship

58   Los Angeles Harbor Light: "Los Angeles Harbor Light," United States Coast Guard Historian's Office, September 10, 2019, https://www.history.uscg.mil/Browse-by-Topic/Assets/Land/Lighthouses-Light-Stations/Article/1956405/los-angeles-harbor-light.

60   extraordinarily long molecules: Christine DeMerchant, "Materials Used for Ropes: High Modulus Polyethelyne HMPE, Dyneema®, Spectra®," Christine DeMerchant.com, https://www.christinedemerchant.com/rope_material_hmpe.html.

60   commercialized until 1990: "Romancing the Thread: The Story of Dyneema," The Dyneema Project, https://www.thedyneemaproject.com/en_gb-old/story-of-dyneema/dyneema.html.

61   more than 9 million in 2019: "Container Statistics," Port of Los Angeles, https://www.portoflosangeles.org/business/statistics/container-statistics.

61   Winston Churchill and Joseph Stalin: "The Battleship of Presidents," Battleship USS *Iowa* Museum, https://www.pacificbattleship.com/the-battleship-of-presidents.

65   murder someone: Robert Escobar, *Saps, Blackjacks and Slungshots: A History of Forgotten Weapons* (Columbus, OH: Gatekeeper Press, 2018).

Chapter 6: Longshoremen Against the Machine

67   throwing a wedding: Alexis Madrigal, *Containers* (podcast), episode 3, "The Ships, the Tugs, and the Port," *Fusion Media Group*, https://www.flexport.com/blog/alexis-madrigal-containers-podcast.

69   six figures: Chris Kirkham and Andrew Khouri, "How Longshoremen Command $100K Salaries in Era of Globalization and Automation," *Los Angeles*

Times, March 2, 2015, https://www.latimes.com/business/la-fi-dockworker-pay-20150301-story.html.

70   more than $95,000 a year: Bill Mongelluzzo, "LA Terminal Automation Delayed, but Not Derailed," JOC.com, February 22, 2019, https://www.joc.com/port-news/port-productivity/la-terminal-automation-delayed-not-derailed_20190222.html.

70   unlimited rights to automate these ports: Bill Mongelluzzo, "ILWU-PMA Contract No Game Changer for West Coast Productivity," JOC.com, May 23, 2015, https://www.joc.com/port-news/longshoreman-labor/international-longshore-and-warehouse-union/ilwu-pma-contract-no-gamechanger-west-coast-productivity_20150523.html.

70   port automation beginning in 1993: Victor M. Sanz, "Welcome to FutureLand." *Volume* 49, October 6, 2016, http://volumeproject.org/welcome-to-futureland.

70   3 percent of the world's shipping terminals are automated: Bill Mongelluzzo, "More North American Port Automation Expected," JOC.com, July 4, 2019, https://www.joc.com/port-news/port-productivity/more-north-american-port-automation-coming-moody%E2%80%99s_20190704.html.

71   force the ports to reopen: David E. Sanger and Steven Greenhouse, "President Invokes Taft-Hartley Act to Open 29 Ports," *New York Times*, October 9, 2002, https://www.nytimes.com/2002/10/09/us/president-invokes-taft-hartley-act-to-open-29-ports.html.

71   Mechanization and Modernization Agreement: Peter Cole, "The Tip of the Spear: How Longshore Workers in the San Francisco Bay Area Survived the Containerization Revolution," *Employee Responsibilities and Rights Journal* 25 (2013): 201–16.

72   500 a day at a single terminal: Hayley Munguia, "This Is What the Automation Deal Means for Workers at Port of LA, Particularly Those at Its Largest Terminal," *Daily Breeze*, July 24, 2019, https://www.dailybreeze.com/2019/07/19/this-is-what-the-automation-deal-means-for-workers-at-port-of-la-particularly-those-at-its-largest-terminal.

72   packed public hearings: Bradley Bermont, "Port of LA Longshoremen Protest Plan to Increase Automation at One Terminal," *Daily Breeze*, January 25, 2019, https://www.dailybreeze.com/2019/01/24/port-of-la-longshoremen-protest-plan-to-increase-automation-at-one-terminal.

72   intercession by L.A. mayor Eric Garcetti: "Garcetti to Mediate Dispute Over Automation at Port of LA," Spectrum News 1, April 23, 2019, https://spectrumnews1.com/ca/la-west/this-month-with-the-mayor/2019/04/23/garcetti-to-mediate-dispute-over-automation-at-port-of-la.

73   withholding authorization for APM to install charging equipment: Dakota Smith, "City Deals Blow to Automation Plan at the Port of L.A. The robots could still be coming," *Los Angeles Times*, June 28, 2019, https://www.latimes.com/local/lanow/la-me-ln-deal-port-dockworkers-automation-ilwu-union-permit-245-maersk-robot-jobs-20190628-story.html.

73   some portion of the workers within an industry: Cole, "The Tip of the Spear."

73   choke points in the global supply chain: Jake Alimahomed-Wilson and Immanuel Ness, eds., *Choke Points: Logistics Workers Disrupting the Global Supply Chain* (London: Pluto Press, 2018).

73   full pensions: Andrea Bernstein, "Why 80K People Applied for 2,400 Positions at LA's Ports," KPCC, June 5, 2017, https://www.scpr.org/news /2017/06/05/72573/why-80k-people-applied-for-2-400-positions-at-la-s.

74   along with globalization: Matthew C. Klein and Michael Pettis, *Trade Wars Are Class Wars: How Rising Inequality Distorts the Global Economy and Threatens International Peace* (New Haven: Yale University Press, 2020).

74   polarization of wealth and wages: Christopher Mims, "Covid-19 Is Dividing the American Worker," *Wall Street Journal*, August 22, 2020, https://www .wsj.com/articles/covid-19-is-dividing-the-american-worker-11598068859.

74   "Hopefully automation kicks in": Margot Roosevelt, "As L.A. Ports Automate, Some Workers Are Cheering on the Robots," *Los Angeles Times*, November 7, 2019, https://www.latimes.com/business/story/2019-11-07/port-automation -dockworkers-vs-truckers.

75   ninety minutes or more: Michael Angell, "Automated Terminals the Fastest Option for Trucking Containers," *American Shipper*, July 22, 2019, https:// www.freightwaves.com/news/automated-terminals-the-fastest-option-in -container-trucking.

75   2030 goal: "Port of Los Angeles Continues Clean Air Progress," Port of Los Angeles, October 1, 2020, https://www.portoflosangeles.org/references /news_100120_air_emissions_report.

76   publicly acknowledged by port operators: Bill Mongelluzzo, "APM Terminal's Automation at LA Foreshadows More," JOC.com, January 25, 2019, https://www.joc.com/port-news/us-ports/port-los-angeles/apm-terminal %E2%80%99s-automation-la-foreshadows-more_20190125.html.

76   has found that it's recessions: Mark Muro, Robert Maxim, and Jacob Whiton, "The Robots Are Ready as the COVID-19 Recession Spreads," *The Avenue* (blog), Brookings Institution, March 24, 2020, https://www.brookings .edu/blog/the-avenue/2020/03/24/the-robots-are-ready-as-the-covid-19 -recession-spreads.

Chapter 7: The Largest Robots on Earth

79   164 feet: "Kalmar RMG Data Sheet," Kalmar Global, https://www.kalmarglobal .com/49c3b0/globalassets/media/11038/11038_RMG-Data-sheet-web -singles.pdf.

79   the Alameda Corridor: "Alameda Corridor Fact Sheet," Alameda Corridor Transportation Authority, http://www.acta.org/projects/projects_completed _alameda_factsheet.asp.

80   globally significant container ports: Jean-Paul Rodrigue and Theo Notteboom, "Port Terminals," from Jean-Paul Rodrigue, *The Geography of Transport Systems* (New York: Routledge, 2020). Retrieved from https://transportgeography .org/?page_id=3235.

81   half of the moves: Fox Chu, Sven Gailus, Lisa Liu, and Liumin Ni, "The Future of Automated Ports," McKinsey & Company, December 4, 2018, https://www.mckinsey.com/industries/travel-logistics-and-transport-infrastructure/our-insights/the-future-of-automated-ports.

82   Copenhagen Telephone Company: Christopher Charles Heyde, "Erlang, Agner Krarup," *Encyclopedia of Mathematics*, https://encyclopediaofmath.org/wiki/Erlang,_Agner_Krarup.

82   change the world: C. C. Heyde, "Agner Krarup Erlang," in *Statisticians of the Centuries*, ed. C. C. Hyde and E. Seneta (New York: Springer-Verlag, 2005), 328–30.

83   "genetic algorithms": Erhan Kozan and Peter Preston, "Genetic Algorithms to Schedule Container Transfers at Multimodal Terminals," *International Transactions in Operational Research* 6, no. 3 (1999): 311–29.

83   artificial neural networks: Ioanna Kourounioti, Amalia Polydoropoulou, and Christos Tsiklidis, "Development of Models Predicting Dwell Time of Import Containers in Port Container Terminals—an Artificial Neural Networks Application," *Transportation Research Procedia* 14 (2016): 243–52.

Chapter 8: The Little-Known, Rarely Understood Organizing Principle of Modern Work

87   Interstate Commerce Commission: Robert Kanigel, *The One Best Way: Frederick Winslow Taylor and the Enigma of Efficiency* (Cambridge, MA: MIT Press, 2005), pp. 433–34.

87   Lenin wrote that scientific management: Zenovia A. Sochor, "Soviet Taylorism Revisited," *Soviet Studies* 33, no. 2 (1981): 246–264, http://www.jstor.org/stable/151338.

88   held the kaiser at bay: Francesca Tesi, "Michelin et le Taylorisme," *Histoire, Économie & Société* 27, no. 3 (2008): 111–26.

88   winning the Second World War: Peter F. Drucker, "Management and the World's Work," *Harvard Business Review*, September 1988, https://hbr.org/1988/09/management-and-the-worlds-work.

88   to factories overseas: Daniel Nelson, ed., *A Mental Revolution: Scientific Management Since Taylor* (Columbus: Ohio State University Press, 1992).

91   almost exclusively women: Joshua B. Freeman, *Behemoth: A History of the Factory and the Making of the Modern World* (New York: W.W. Norton, 2018).

91   two out of three women: Joyce Burnette, "Women Workers in the British Industrial Revolution," EH.Net Encyclopedia, ed. Robert Whaples, March 26, 2008, https://eh.net/encyclopedia/women-workers-in-the-british-industrial-revolution-2.

Chapter 9: How a Management Philosophy Became Our Way of Life

93   ardent abolitionist: Robert Kanigel, *The One Best Way: Frederick Winslow Taylor and the Enigma of Efficiency* (Cambridge, MA: MIT Press, 2005), pp. 433–34.

94   fit the first one perfectly: Simon Winchester, *The Perfectionists* (New York: HarperCollins, 2018).

95   psychology of skilled workers: Frederick Winslow Taylor, *The Principles of Scientific Management* (New York: Harper & Brothers, 1911).

96   patented in 1891: Willard Le Grand Bundy, workman's time recorder, US Patent 452894A, patented May 26, 1891.

96   Waltham Watch Company: "Chronodrometer or Horse Timing Watch," National Museum of American History, https://americanhistory.si.edu/collections /search/object/nmah_852606.

97   wrote historian Jill Lepore: Jill Lepore, "Not So Fast," *The New Yorker*, October 12, 2009, https://www.newyorker.com/magazine/2009/10/12/not-so-fast.

98   the name of their philosophy: Horace Bookwalter Drury, *Scientific Management: A History and Criticism* (New York: Columbia University, 1918).

98   "This method is un-American": Drury, *Scientific Management*.

98   Daniel Nelson wrote: Daniel Nelson, *Frederick W. Taylor and the Rise of Scientific Management* (Madison: University of Wisconsin Press, 1980).

100   "multiple discovery": William F. Ogburn and Dorothy Thomas, "Are Inventions Inevitable? A Note on Social Evolution," *Political Science Quarterly* 37, no. 1 (1922): 83–98.

101   facilities in Chicago: Boris Emmet and John E. Jeuck, *Catalogues and Counters: A History of Sears, Roebuck & Company* (Chicago: University of Chicago Press, 1965).

102   "French military officers had recognized the potential of scientific management": Nelson, *Frederick W. Taylor and the Rise of Scientific Management*.

104   wrote in 1955: Lillian M. Gilbreth, *Management in the Home: Happier Living Through Saving Time and Energy* (New York: Dodd, Mead & Co., 1955).

104   beholding her work: Lepore, "Not So Fast."

104   did not cook: Alexandra Lange, "The Woman Who Invented the Kitchen," *Slate*, October 25, 2012, https://slate.com/human-interest/2012/10/lillian -gilbreths-kitchen-practical-how-it-reinvented-the-modern-kitchen.html.

104   a writer described: Blanche Halbert, ed., *The Better Homes Manual* (Chicago: University of Chicago Press, 1931).

105   their enhanced abilities: "Dr. Gilbreth's Kitchen," *On Time* (blog), National Museum of American History, https://americanhistory.si.edu/ontime/saving /kitchen.html.

105   wrote historian Ruth Schwartz Cowan: Ruth Schwartz Cowan, *More Work for Mother: The Ironies of Household Technology from the Open Hearth to the Microwave* (New York: Basic Books, 1995).

105   missed the point entirely: Frederick Winslow Taylor, "An Answer to the Criticism," *The American Magazine* 72 (1911).

Chapter 10: Rime of the Long-Haul Trucker

108   every year: Paul Schenck, "Top-25 Trailer Output 2019," Trailer Body Builders, March 13, 2020, https://www.trailer-bodybuilders.com/trailer-output /article/21126082/top25-trailer-output-2019.

108 almost all of them in China: "Coronavirus Disease 2019 (COVID-19) Situation Report—43," World Health Organization, March 3, 2020, https://www.who.int/docs/default-source/coronaviruse/situation-reports/20200303-sitrep-43-covid-19.pdf.

108 "risk to most Americans is low": "An Emerging Disease Threat: How the U.S. Is Responding to COVID-19, the Novel Coronavirus," Centers for Disease Control and Prevention, March 3, 2020, https://www.cdc.gov/washington/testimony/2020/t20200303.htm.

108 effective against Covid-19: "List N: Disinfectants for Coronavirus (COVID-19)," United States Environmental Protection Agency, https://www.epa.gov/pesticide-registration/list-n-disinfectants-coronavirus-covid-19.

110 begun by Richard Nixon: Donald V. Harper, "The Federal Motor Carrier Act of 1980: Review and Analysis," *Transportation Journal* 20, no. 2 (1980): 5–33.

110 "increased job opportunities": "Motor Carrier Act of 1980 Statement on Signing S. 2245 into Law," American Presidency Project, https://www.presidency.ucsb.edu/documents/motor-carrier-act-1980-statement-signing-s-2245-into-law.

110 $120,000 in 2020 dollars: James Jaillet, "Trucker Pay Has Plummeted in the Last 30 Years, Analyst Says," *Overdrive*, March 4, 2016, https://www.overdriveonline.com/trucker-pay-has-plummeted-in-the-last-30-years-analyst-says.

110 $45,000 a year: "Heavy and Tractor-Trailer Truck Drivers," U.S. Bureau of Labor Statistics Occupational Outlook Handbook, September 16, 2020, https://www.bls.gov/ooh/transportation-and-material-moving/heavy-and-tractor-trailer-truck-drivers.htm.

111 dire consequences for us all: Bob Costello and Alan Karickhoff. "ATA Releases Updated Driver Shortage Report and Forecast." American Trucking Associations, July 23, 2020. https://www.trucking.org/news-insights/ata-releases-updated-driver-shortage-report-and-forecast.

111 2019 paper: Stephen V. Burks and Kristen Monaco, "Is the U.S. Labor Market for Truck Drivers Broken?" *Monthly Labor Review* (U.S. Bureau of Labor Statistics), March 2019, https://www.bls.gov/opub/mlr/2019/article/is-the-us-labor-market-for-truck-drivers-broken.htm.

112 first ninety days: Roger Gilroy, "Stay Metrics Identifies Why Drivers Leave So Soon," *Transport Topics*, October 7, 2019, https://www.ttnews.com/articles/stay-metrics-identifies-why-drivers-leave-so-soon.

112 $68,000 a year: "Fleet Owner 500: Top Private Fleets of 2019," *FleetOwner*, April 15, 2019, https://www.fleetowner.com/truck-stats/fleet-owner-500/article/21703705/fleet-owner-500-top-private-fleets-of-2019.

112 "throwaway" person: Trip Gabriel, "Alone on the Open Road: Truckers Feel Like 'Throwaway People,'" *New York Times*, May 22, 2017, https://www.nytimes.com/2017/05/22/us/trucking-jobs.html.

113 fast-food outlets: Emily Guendelsberger, *On the Clock: What Low-Wage Work Did to Me and How It Drives America Insane* (New York: Little, Brown and Company, 2019).

113 Rose George: Rose George, *Ninety Percent of Everything: Inside Shipping, the*

Invisible Industry That Puts Clothes on Your Back, Gas in Your Car, and Food on Your Plate (New York: Picador, 2014).

114 truck drivers are men: Cristina Roca and Dieter Holger, "Drawn by the Salary, Women Flock to Trucking," *Wall Street Journal*, October 14, 2019, https://www.wsj.com/articles/drawn-by-the-salary-women-flock-to-trucking-11571045406.

117 get more rest: Michael Belzer, "Truck Drivers Are Overtired, Overworked and Underpaid," *The Conversation*, July 25, 2018, https://theconversation.com/truck-drivers-are-overtired-overworked-and-underpaid-100218.

118 up 23 percent from 2010: "Large Truck and Bus Crash Facts 2018," Federal Motor Carrier Safety Administration, October 2, 2020, https://www.fmcsa.dot.gov/safety/data-and-statistics/large-truck-and-bus-crash-facts-2018.

118 beginning of those conflicts: "US & Allied Killed," Watson Institute for International and Public Affairs Costs of War Project (Brown University), January 2020, https://watson.brown.edu/costsofwar/costs/human/military/killed.

Chapter 11: 100 Percent of Everything

121 2018 study of truckers: Adam Hege, Michael K. Lemke, Yorghos Apostolopoulos, and Sevil Sönmez, "Occupational Health Disparities Among U.S. Long-Haul Truck Drivers: The Influence of Work Organization and Sleep on Cardiovascular and Metabolic Disease Risk," *PloS ONE* 13, no. 11 (2018), https://journals.plos.org/plosone/article?id=10.1371/journal.pone.0207322.

122 "AirSpace": Kyle Chayka, "Welcome to AirSpace," *The Verge*, August 3, 2016, https://www.theverge.com/2016/8/3/12325104/airbnb-aesthetic-global-minimalism-startup-gentrification.

124 thirty years of political and legal wrangling: Hannah Steffensen, "A Timeline of the ELD Mandate: History & Important Dates," GPS Trackit, May 3, 2017, https://gpstrackit.com/blog/a-timeline-of-the-eld-mandate-history-and-important-dates.

124 no effect on the number of crashes: Alex Scott, Andrew Balthrop, and Jason Miller, "Did the Electronic Logging Device Mandate Reduce Accidents?" January 11, 2019. Available at SSRN, https://ssrn.com/abstract=3314308.

126 half of America's 5,400 Walmarts: Ashley Coker, "Walmart Faces $50k Fine for Allowing Truck Parking at Illinois Store," *Freight Waves*, March 26, 2019, https://www.freightwaves.com/news/walmart-faces-50k-fine-for-allowing-truck-parking-at-illinois-store.

Chapter 12: How "Hitler's Highway" Became America's Circulatory System

128 Adolf Hitler: Tom Lewis, *Divided Highways: Building the Interstate Highways, Transforming American Life* (Ithaca, NY: Cornell University Press, 2013).

129 "draped with national standards": Dayton Duncan and Ken Burns, *Horatio's Drive: America's First Road Trip* (New York: Alfred A. Knopf, 2003). Excerpted at https://www.pbs.org/kenburns/horatios-drive/documents.

129   early cars and trucks: United States Department of Transportation, Federal Highway Administration, *America's Highways, 1776–1976: A History of the Federal-Aid Program* (Washington, D.C.: U.S. Government Printing Office, 1977).

130   2 million trucks: Athel F. Denham, *20 Years' Progress in Commercial Motor Vehicles* (Detroit: Automotive Council for War Production, 1943).

130   the "townless highway": Benton MacKaye and Lewis Mumford, "Townless Highways for the Motorist," *Harper's*, August 1931.

131   consolidation of power: Eric Jaffe, "How Highway Construction Helped Hitler Rise to Power," *Bloomberg CityLab*, June 6, 2014, https://www.bloomberg .com/news/articles/2014-06-06/how-highway-construction-helped-hitler -rise-to-power.

131   Autobahn was under construction: Nico Voigtlaender and Hans-Joachim Voth, "Highway to Hitler" (working paper, National Bureau of Economic Research, January 2021). Available at https://www.nber.org/system/files /working_papers/w20150/w20150.pdf.

131   By 1941, when construction was halted: "The Reichsautobahnen," *Highway History* (blog), Federal Highway Administration, June 27, 2017, https:// www.fhwa.dot.gov/infrastructure/reichs.cfm.

132   concerned chiefly with the movement of automobiles: Richard F. Weingroff, "Edward M. Bassett the Man Who Gave Us 'Freeway.'" *Highway History* (blog), Federal Highway Administration, June 27, 2017, https://www.fhwa .dot.gov/infrastructure/freeway.cfm.

132   "ribbons across the land": Richard Weingroff, "Zero Milestone— Washington, DC," *Highway History* (blog), Federal Highway Administration, June 27, 2017, https://www.fhwa.dot.gov/infrastructure/zero.cfm.

132   164,000 miles of highway: "Highway Finance Data Collection," Federal Highway Administration, November 7, 2014, https://www.fhwa.dot.gov /policyinformation/pubs/hf/pl11028/chapter1.cfm.

132   1.5 billion metric tons: "Materials in Use in U.S. Interstate Highways," United States Geological Survey, October 2006, https://pubs.usgs.gov /fs/2006/3127/2006-3127.pdf.

132   3 *trillion* miles: "Strong Economy Has Americans Driving More than Ever Before," Federal Highway Administration, March 21, 2019, https://www .fhwa.dot.gov/pressroom/fhwa1905.cfm.

132   12 billion tons of freight: "Trucking Moved 11.84 Billion Tons of Freight in 2019," American Trucking Associations, July 13, 2020, https://www.trucking .org/news-insights/trucking-moved-1184-billion-tons-freight-2019.

136   single-digit percentage of the overall market: "Fleet Owner 500: Top Private Fleets of 2019," *FleetOwner*, April 15, 2019, https://www.fleetowner.com /truck-stats/fleet-owner-500/article/21703705/fleet-owner-500-top-private -fleets-of-2019.

138   $3.6 billion into trucking: Cyndia Zwahlen, "Freight Tech VC on Track to Top 2018's Record $3.6 billion," Trucks.com, April 29, 2019, https://www .trucks.com/2019/04/29/freight-tech-vc-top-record-3-6-billion.

Chapter 13: The Future of Trucking

142 brittle and shallow: Jason Pontin, "Greedy, Brittle, Opaque, and Shallow: The Downsides to Deep Learning," *Wired*, February 2, 2018, https://www .wired.com/story/greedy-brittle-opaque-and-shallow-the-downsides-to -deep-learning.

142 a map unlike any created before it: Christopher Mims, "The Key to Autonomous Driving? An Impossibly Perfect Map," *Wall Street Journal*, October 11, 2018, https://www.wsj.com/articles/the-key-to-autonomous-driving-an-impossibly -perfect-map-1539259260.

144 launched in New Mexico in 1932: "Dr. Robert H. Goddard, American Rocketry Pioneer," NASA Goddard Space Flight Center, February 11, 2015, https://www.nasa.gov/centers/goddard/about/history/dr_goddard .html.

144 Operation Paperclip: Annie Jacobsen, *Operation Paperclip: The Secret Intelligence Program That Brought Nazi Scientists to America* (New York: Little, Brown and Company, 2015).

144 getting humans to the moon: "The Inertial Measurement Unit: Mechanical Engineering Wizardry," Hack the Moon, https://wehackthemoon.com/tech /inertial-measurement-unit-mechanical-engineering-wizardry.

146 about two minutes: Richard W. Pogge, "Real-World Relativity: The GPS Navigation System," Ohio State University, March 11, 2017, http://www .astronomy.ohio-state.edu/~pogge/Ast162/Unit5/gps.html.

151 Bayesian analysis: Derek Muller (Veritasium), "The Bayesian Trap," YouTube, April 5, 2017, https://www.youtube.com/watch?v=R13BD8qKeTg.

153 Moore's law has slowed: Christopher Mims, "Huang's Law Is the New Moore's Law, and Explains Why Nvidia Wants Arm," *Wall Street Journal*, September 19, 2020, https://www.wsj.com/articles/huangs-law-is-the-new -moores-law-and-explains-why-nvidia-wants-arm-11600488001.

157 local minimum wage happens to be: Greg Bensinger, "Uber's Driver Dilemma: Fare Hikes and Cuts Don't Change Pay," *Wall Street Journal*, November 12, 2017, https://www.wsj.com/articles/ubers-driver-dilemma-fare-hikes-and -cuts-dont-change-pay-1510491602.

Chapter 14: What Actually Happens Inside Amazon's Warehouses

160 2 billion totes: Brian Heater, "Amazon Says It Has Deployed More than 200,000 Robotic Drives Globally," *TechCrunch*, June 5, 2019, https://techcrunch .com/2019/06/05/amazon-says-it-has-deployed-more-than-200000-robotic -drives-globally.

160 "300,000 associates": "What Robots Do (and Don't Do) at Amazon Fulfilment Centres," About Amazon, https://www.aboutamazon.co.uk/amazon-fulfilment /what-robots-do-and-dont-do-at-amazon-fulfilment-centres.

160 200 semi-automated "fulfillment centers": "Amazon Supply Chain and Ful-

fillment Center Network," MWPVL International, https://mwpvl.com/html/amazon_com.html.

160 200,000 of these robots: Heater, "Amazon Says It Has Deployed More than 200,000."

162 bargain-hunting resellers: Rachel Siegel, "'Flesh and Blood Robots for Amazon': They Raid Clearance Aisles and Resell It All Online for a Profit," *Washington Post*, February 8, 2019, https://www.washingtonpost.com/business/economy/flesh-and-blood-robots-for-amazon-they-raid-clearance-aisles-and-resell-it-all-online-for-a-profit/2019/02/08/f71bff72-2a60-11e9-984d-9b8fba003e81_story.html.

162 fish items out of the trash: Khadeeja Safdar, Shane Shifflett, and Denise Blostein, "You Might Be Buying Trash on Amazon—Literally," *Wall Street Journal*, December 18, 2019, https://www.wsj.com/articles/you-might-be-buying-trash-on-amazonliterally-11576599910.

162 AWS encompasses: "Cloud Products," Amazon Web Services, https://aws.amazon.com/products.

164 You've got to be crazy: Jennifer Alsever, "Robot Workers Take Over Warehouses," CNN, November 9, 2011, https://money.cnn.com/2011/11/09/smallbusiness/kiva_robots/index.htm.

167 pioneered by Walmart: Charles Fishman, *The Wal-Mart Effect* (London: Penguin Books, 2006).

167 16 percent of all retail sales: "Commerce Retail Sales as a Percent of Total Sales," Federal Reserve Bank of St. Louis, November 19, 2020, https://fred.stlouisfed.org/series/ECOMPCTSA.

167 quarter of all retail sales: Derek Mahlburg, "What's Behind the US E-commerce Statistics?" RISI Fastmarkets, June 2019, https://insights.risiinfo.com/e-commerce_statistics_June_2019/index.html.

167 Bill Gross: Bill Gross, "The Single Biggest Reason Why Start-ups Succeed," TED2015, March 2015, https://www.ted.com/talks/bill_gross_the_single_biggest_reason_why_start_ups_succeed.

168 in talks with Jeff Wilke: Gigi Marino, "Geeking Out at Amazon," *MIT Technology Review*, October 15, 2007, https://www.technologyreview.com/2007/10/15/223422/geeking-out-at-amazon.

169 "robot arms race": Kim Bhasin and Patrick Clark, "How Amazon Triggered a Robot Arms Race," *Bloomberg News*, June 29, 2016, https://www.bloomberg.com/news/articles/2016-06-29/how-amazon-triggered-a-robot-arms-race.

169 "Amazon Prime effect": Christopher Mims, "The Prime Effect: How Amazon's Two-Day Shipping Is Disrupting Retail," *Wall Street Journal*, September 20, 2018, https://www.wsj.com/articles/the-prime-effect-how-amazons-2-day-shipping-is-disrupting-retail-1537448425.

171 cover of *Wired* magazine: Jessica Bruder, "Meet the Immigrants Who Took On Amazon," *Wired*, November 12, 2019, https://www.wired.com/story/meet-the-immigrants-who-took-on-amazon.

Chapter 15: The Unbearable Complexity of Robotic Warehousing

179   350 pounds: *Staten Island Advance*, "A Peek Inside the New Amazon Fulfillment Center: The Robots," YouTube, June 19, 2018, https://www.youtube.com /watch?v=lu9aVOrz4HU.

185   you get this efficiency for "free": Christopher Mims, "How Robots and Drones Will Change Retail Forever," *Wall Street Journal*, October 15, 2018, https://www.wsj.com/articles/how-robots-and-drones-will-change-retail -forever-1539604800.

186   350 at minimum: Lydia DePillis and Richa Naik, "Amazon's Incredible, Vanishing Cardboard Box," CNN, July 16, 2019, https://www.cnn.com /2019/07/16/business/amazon-cardboard-box-prime-day/index.html.

188   The company doesn't talk publicly about AFE: John Burgett, "AFE (Amazon Fulfillment Engine)," Amazon Emancipatory, 2016, https://amazonemancipatory .com/afe-amazon-fulfillment-engine.

188   its inner workings: Rodney Galeano, "OB AFE 3.0 Managers Guide," Quizlet, https://quizlet.com/216728746/ob-afe-30-managers-guide-flash-cards.

191   "AFE remains brutal": Burgett, "AFE (Amazon Fulfillment Engine)."

195   how long such orders will last: "Coronavirus Updates from March 20, 2020," CBS News, March 21, 2020, https://www.cbsnews.com/live-updates /coronavirus-disease-covid-19-latest-news-2020-03-20.

195   panic buying of necessities and staples: Corina Knoll, "Panicked Shoppers Empty Shelves as Coronavirus Anxiety Rises," *New York Times*, March 13, 2020, https://www.nytimes.com/2020/03/13/nyregion/coronavirus-panic -buying.html.

195   sold out of toilet paper: Anthony Cuthbertson, "Coronavirus: Amazon Sells Out of Toilet Paper and Other Essentials amid Panic Buying," *Independent* (U.K.), March 16, 2020, https://www.independent.co.uk/life-style/gadgets-and -tech/news/coronavirus-amazon-panic-buying-shortages-stockpile-covid -19-a9404631.html.

195   A study published by a pair of economists: Michael Keane and Timothy Neal, "Consumer Panic in the COVID-19 Pandemic," *Journal of Econometrics* 220, no 1 (January 2021): 86–105.

Chapter 16: Bezosism

198   "algorithmic despotism": Alina Selyukh, "At the Mercy of an App: Workers Feel the Instacart Squeeze," NPR, November 25, 2019, https://www.npr .org/2019/11/25/778546287/at-the-mercy-of-an-app-workers-feel-the -instacart-squeeze.

201   Amazon's global footprint: "Amazon Supply Chain and Fulfillment Center Network," MWPVL International, https://www.mwpvl.com/html/amazon _com.html.

204   "rank and yank": Max Nisen, "A Lawsuit Claims Microsoft's Infamous Stack Rankings Made Things Worse for Women," *Quartz*, September 15, 2015,

https://qz.com/504507/a-lawsuit-claims-microsofts-infamous-stack-rankings-made-things-worse-for-women.

204 abandoned it: Shira Ovide and Rachel Feintzeig, "Microsoft Abandons 'Stack Ranking' of Employees," *Wall Street Journal*, November 12, 2013, https://www.wsj.com/articles/SB1000142405270230346000457919395198766 16572.

208 database of the injury rates: Will Evans, "Ruthless Quotas at Amazon Are Maiming Employees," *The Atlantic*, November 25, 2019, https://www.theatlantic.com/technology/archive/2019/11/amazon-warehouse-reports-show-worker-injuries/602530.

208 like a robot: Michael Sainato, "Amazon Workers Condemn Unsafe, Grueling Conditions at Warehouse," *The Guardian* (U.K.), February 5, 2020, https://www.theguardian.com/technology/2020/feb/05/amazon-workers-protest-unsafe-grueling-conditions-warehouse.

210 jobs of middle managers: Christopher Mims, "Data Is the New Middle Manager," *Wall Street Journal*, April 19, 2015, https://www.wsj.com/articles/data-is-the-new-middle-manager-1429478017.

211 force for good or ill: Ben Thompson, "The China Cultural Clash," *Stratechery*, July 25, 2020, https://stratechery.com/2019/the-china-cultural-clash.

211 John Maynard Keynes: Elizabeth Kolbert, "No Time," *The New Yorker*, May 19, 2014, https://www.newyorker.com/magazine/2014/05/26/no-time.

211 six and a half hours a day: Derek Thompson, "Workism Is Making Americans Miserable," *The Atlantic*, February 24, 2019, https://www.theatlantic.com/ideas/archive/2019/02/religion-workism-making-americans-miserable/583441.

211 shiftless men: Jillian Berman, "Women's Unpaid Work Is the Backbone of the American Economy," MarketWatch, April 15, 2018, https://www.marketwatch.com/story/this-is-how-much-more-unpaid-work-women-do-than-men-2017-03-07.

211 don't eliminate jobs: Christopher Mims, "Robot Reality Check: They Create Wealth—and Jobs," *Wall Street Journal*, November 30, 2018, https://www.wsj.com/articles/robot-reality-check-they-create-wealthand-jobs-1543500001.

212 As Amazon itself puts it: "What Robots Do (and Don't Do) at Amazon Fulfilment Centres," About Amazon, https://www.aboutamazon.co.uk/amazon-fulfilment/what-robots-do-and-dont-do-at-amazon-fulfilment-centres.

212 huge demand for enslaved persons: Joan Brodsky Schur, "Eli Whitney's Patent for the Cotton Gin," National Archives and Records Administration, September 23, 2016, https://www.archives.gov/education/lessons/cotton-gin-patent.

213 Taylor made the same error: Frederick Winslow Taylor, *The Principles of Scientific Management* (New York: Harper & Brothers, 1911).

213 This trend was later confirmed: Will Evans, "Leaked Documents Show How Amazon Misled the Public About Warehouse Safety Issues," *PBS NewsHour*, October 13, 2020, https://www.pbs.org/newshour/show/leaked-documents-show-how-amazon-misled-the-public-about-warehouse-safety-issues.

213   "Before robots, it was still tough": Will Evans, "Behind the Smiles," *Reveal* (Center for Investigative Reporting), November 25, 2019, https://revealnews .org/article/behind-the-smiles.

215   millions of human minds: Arthur Herman, *Freedom's Forge: How American Business Produced Victory in World War II* (New York: Random House, 2013).

215   "cycle time": James P. Womack, Daniel T. Jones, and Daniel Roos, *The Machine That Changed the World* (New York: HarperCollins, 1991).

216   noblesse oblige: Abha Bhattarai, "Amazon Boosts Minimum Wage to $15 for All Workers Following Criticism," *Washington Post*, October 2, 2018, https:// www.washingtonpost.com/business/2018/10/02/amazon-announces-it -will-boost-minimum-wage-all-workers-after-facing-criticism.

216   850,000 applications: Nat Levy, "Amazon Got 850K Job Applications in One Month After Announcing a $15 Minimum Wage," *GeekWire*, January 31, 2019, https://www.geekwire.com/2019/amazon-got-850k-job-applications -announcing-15-minimum-wage.

218   high levels of unemployment: Alana Semuels, "What Amazon Does to Poor Cities," *The Atlantic*, February 1, 2018, https://www.theatlantic.com/business /archive/2018/02/amazon-warehouses-poor-cities/552020.

220   "upskill" 100,000 of its employees: "Amazon Pledges to Upskill 100,000 U.S. Employees for In-Demand Jobs by 2025," press release, About Amazon, July 11, 2019, https://press.aboutamazon.com/news-releases/news-release -details/amazon-pledges-upskill-100000-us-employees-demand-jobs-2025.

Chapter 17: From Japan with Love

222   Robert McNamara's approach: Errol Morris, *The Fog of War: Eleven Lessons from the Life of Robert S. McNamara* (Sony Pictures Classics, 2003).

223   fortunes of GE declined: Oliver Staley, "Whatever Happened to Six Sigma?" *Quartz*, September 3, 2019, https://qz.com/work/1635960/whatever-happened -to-six-sigma.

224   operation dubbed Save Santa: Gretchen Gavett, "How One Bad Thanksgiving Shaped Amazon," *Harvard Business Review*, August 7, 2014, https://hbr.org /2013/11/how-one-bad-thanksgiving-shaped-amazon.

225   never make money on it: Brad Stone, *The Everything Store: Jeff Bezos and the Age of Amazon* (New York: Little, Brown and Company, 2013).

232   "so as to empower or to enslave": Paul Adler, "Democratic Taylorism: The Toyota Production System at NUMMI," In "Lean Work: Empowerment and Exploitation in the Global Auto Industry."

Chapter 18: How Warehouse Work Injures

233   the effects of automation on workers at Amazon: Beth Gutelius and Nik Theodore, "The Future of Warehouse Work: Technological Change in the U.S. Logistics Industry," UC Berkeley Center for Labor Research and Education, October 22, 2019, https://laborcenter.berkeley.edu/future-of-warehouse-work.

235 workforce as a whole: Elka Torpey, "Employment Growth and Wages in E-commerce: Career Outlook," U.S. Bureau of Labor Statistics, December 2018, https://www.bls.gov/careeroutlook/2018/article/e-commerce-growth .htm?view_full.

235 May 2019 report: Martha Ross, "Most Out-of-Work Young Adults Face Bleak Job Prospects," *The Avenue* (blog), Brookings Institution, May 16, 2019, https://www.brookings.edu/blog/the-avenue/2019/05/15/most-out-of -work-young-adults-face-bleak-job-prospects.

237 Beth Galetti: Harry McCracken, "Meet the Woman Behind Amazon's Explosive Growth," *Fast Company*, April 11, 2019, https://www.fastcompany .com/90325624/yes-amazon-has-an-hr-chief-meet-beth-galetti.

237 270,000 such injuries: "Occupational Injuries and Illnesses Resulting in Musculoskeletal Disorders (MSDs)," U.S. Bureau of Labor Statistics, May 1, 2020, https://www.bls.gov/iif/oshwc/case/msds.htm.

237 One in two American adults: American Academy of Orthopaedic Surgeons, "One in Two Americans Have a Musculoskeletal Condition," ScienceDaily, March 1, 2016, https://www.sciencedaily.com/releases/2016/03/160301114116.htm.

238 more dangerous than coal mining: Gutelius, and Theodore, "The Future of Warehouse Work."

238 camera-based systems: Behnoosh Parsa et al., "Toward Ergonomic Risk Prediction via Segmentation of Indoor Object Manipulation Actions Using Spatiotemporal Convolutional Networks," *IEEE Robotics and Automation Letters* 4 (2019): 3153–3160.

238 recover after performing it: Mark Middlesworth, "A Step-by-Step Guide to the REBA Assessment Tool," ErgoPlus, https://ergo-plus.com/reba-assessment -tool-guide.

239 3 million Americans who are seriously injured: Talia Buford and Maryam Jameel, "Workplace Injury, Illness Costs Being Foisted on Workers, Government, OSHA's Michaels Says," Center for Public Integrity, March 4, 2015, https://publicintegrity.org/inequality-poverty-opportunity/workers -rights/workplace-injury-illness-costs-being-foisted-on-workers -government-oshas-michaels-says.

240 habitually underreport: Steven Greenhouse, "Work-Related Injuries Underreported," *New York Times*, November 16, 2009, https://www.nytimes.com /2009/11/17/us/17osha.html.

Chapter 19: Amazon's Employee-Lite Endgame

242 "gale of creative destruction": Joseph A. Schumpeter, *Capitalism, Socialism, and Democracy*, 3rd ed. (New York: Harper & Brothers, 1942).

246 "grasping is going to be a solved problem": Jeffrey Dastin, "Amazon's Bezos Says Robotic Hands Will Be Ready for Commercial Use in Next 10 Years," Reuters, June 6, 2019, https://www.reuters.com/article/us-amazon -com-conference/amazons-bezos-says-robotic-hands-will-be-ready-for -commercial-use-in-next-10-years-idUSKCN1T72JB.

246 robotics innovation hub: Ron Miller, "Amazon Is Planning a $40M Robotics Hub Near Boston," *TechCrunch*, November 6, 2019, https://techcrunch.com/2019/11/06/amazon-is-planning-a-40m-robotics-hub-near-boston.

247 Unimate: Rebecca J. Rosen, "Unimate: The Story of George Devol and the First Robotic Arm," *The Atlantic*, August 16, 2011, https://www.theatlantic.com/technology/archive/2011/08/unimate-the-story-of-george-devol-and-the-first-robotic-arm/243716.

250 created more jobs: Greg Ip, "Workers: Fear Not the Robot Apocalypse," *Wall Street Journal*, September 5, 2017, https://www.wsj.com/articles/workers-fear-not-the-robot-apocalypse-1504631505.

Chapter 20: The Middle Mile

252 founded in 1907: "Our History," UPS, https://stories.ups.com/upsstories/us/en/about-us/our-history.html.

252 founded in 1971: "FedEx History," FedEx, https://www.fedex.com/en-us/about/history.html.

252 more packages than FedEx: Nick Statt, "Amazon Is Delivering Half Its Own Packages as It Becomes a Serious Rival to FedEx and UPS," *The Verge*, December 13, 2019, https://www.theverge.com/2019/12/13/21020938/amazon-logistics-prime-air-fedex-ups-package-delivery-more-than-50-percent.

252 1.9 billion packages: "Pitney Bowes Parcel Shipping Index Reports Continued Growth as Global Parcel Volume Exceeds 100 billion for First Time Ever," Pitney Bowes, October 12, 2020, http://news.pb.com/article_display.cfm?article_id=5958.

254 40 million packages per day: "Pitney Bowes Parcel Shipping Index Reports."

255 "major disaster": Julia Hollingsworth et al., "March 22, 2020 Coronavirus News," CNN, March 22, 2020, https://www.cnn.com/world/live-news/coronavirus-outbreak-03-22-20/index.html.

Chapter 21: The Future of Delivery

265 $100 billion: "$9.6 Billion in Investments Spurring Aggressive Expansion of Food Delivery Companies," Frost & Sullivan, October 25, 2019, https://ww2.frost.com/news/press-releases/9-6-billion-in-investments-spurring-aggressive-expansion-of-food-delivery-companies.

267 nine cameras: Christopher Mims, "The Scramble for Delivery Robots Is On and Startups Can Barely Keep Up," *Wall Street Journal*, April 25, 2020, https://www.wsj.com/articles/the-scramble-for-delivery-robots-is-on-and-startups-can-barely-keep-up-11587787199.

Chapter 22: The Last Mile

272 "enclothed cognition": Hajo Adam and Adam D. Galinsky, "Enclothed Cognition," *Journal of Experimental Social Psychology* 48, no. 4 (2012): 918–25.

273 destroying their body: Paul Sawers, "UPS Will Now Use Dynamic Routing to Get Parcels to You on Time," *Venture Beat*, January 29, 2020, https://venturebeat.com/2020/01/29/ups-will-now-use-dynamic-routing-to-get-parcels-to-you-on-time.

275 died at the scene: Ken Bensinger et al., "The Fast Mile," *BuzzFeed News*, December 24, 2019, https://www.buzzfeednews.com/article/kenbensinger/amazons-race-to-build-a-fast-delivery-network-the-human.

276 Vice: Lauren Kaori Gurley, "Amazon Delivery Drivers Are Overwhelmed and Overworked by Covid-19 Surge," *Motherboard* (blog), Vice, July 1, 2020, https://www.vice.com/en/article/m7j7mb/amazon-delivery-drivers-are-overwhelmed-and-overworked-by-covid-19-surge.

276 *New York Times*: Patricia Callahan, "Amazon Pushes Fast Shipping but Avoids Responsibility for the Human Cost," *New York Times*, September 5, 2019, https://www.nytimes.com/2019/09/05/us/amazon-delivery-drivers-accidents.html.

276 NBC News: David Ingram and Jo Ling Kent, "Inside Amazon's Delivery Push: Employees and Drivers Say an Overworked System Is Lax on Safety as Packages Pile Up," NBC News, November 27, 2019, https://www.nbcnews.com/tech/tech-news/inside-amazon-s-delivery-push-employees-drivers-say-overworked-system-n1087661.

276 system of incentives: Joe Allen, *The Package King: A Rank-and-File History of UPS* (Chicago: Haymarket Books, 2020).

276 "peak on top of peak": Matt Leonard, "FedEx Readies for 'Peak on Top of Peak' with 4 Tech Investments," *Supply Chain Dive*, October 30, 2020, https://www.supplychaindive.com/news/4-takeaways-fedex-peak-innovation-showcase/588102.

277 goes back to the 1930s: Greg Niemann, *Big Brown: The Untold Story of UPS* (San Francisco: Jossey-Bass, 2007).

278 level not seen since the Great Depression: "Union Membership (Annual) News Release," U.S. Bureau of Labor Statistics, January 22, 2021, https://www.bls.gov/news.release/union2.htm.

278 David Weil, a professor at Brandeis: Caroline O'Donovan and Ken Bensinger, "The Cost of Next-Day Delivery," *BuzzFeed News*, September 6, 2019, https://www.buzzfeednews.com/article/carolineodonovan/amazon-next-day-delivery-deaths.

279 800 different companies: Spencer Soper and Matt Day, "Amazon Axes Delivery Partners in U.S.; Hundreds of Jobs Cut," *Bloomberg News*, February 14, 2020, https://www.bloomberg.com/news/articles/2020-02-14/amazon-ends-ties-to-delivery-partner-erasing-hundreds-of-jobs.

280 accusing a company of violating labor laws: Noam Scheiber, "Labor Department Curbs Announcements of Company Violations," *New York Times*, October 23, 2020, https://www.nytimes.com/2020/10/23/business/economy/labor-department-memo.html.

280 outmoded legacy systems: Paul Ziobro, "UPS's $20 Billion Problem: Operations Stuck in the 20th Century," *Wall Street Journal*, June 15, 2018, https://

www.wsj.com/articles/upss-20-billion-problem-operations-stuck-in-the-20th-century-1529072397.

281   chosen a different route: Gordon L. Weil, *Sears, Roebuck, U.S.A.: The Great American Catalog Store and How It Grew* (New York: Jove Publications, 1979).

281   $30 million: Steven Rosenbush and Laura Stevens, "At UPS, the Algorithm Is the Driver," *Wall Street Journal*, February 16, 2015, https://www.wsj.com/articles/at-ups-the-algorithm-is-the-driver-1424136536.

281   5.5 billion: UPS, 2020 10-K form, United States Security and Exchange Commission, https://www.sec.gov/Archives/edgar/data/1090727/000109072714000009/ups-12312013x10k.htm#:~:text=In%202013%2C%20we%20delivered%20an,total%20of%204.3%20billion%20packages.

282   Clay Mathematics Institute: Ethan Siegel, "This 90 Year Old Math Problem Shows Why We Need Quantum Computers," *Forbes*, May 28, 2020, https://www.forbes.com/sites/startswithabang/2020/05/28/this-90-year-old-math-problem-shows-why-we-need-quantum-computers/?sh=6080e2c01c5d.

283   plotting a route for a driver: Tom Vanderbilt, "Unhappy Truckers and Other Algorithmic Problems," *Nautilus*, July 18, 2013, http://nautil.us/issue/3/in-transit/unhappy-truckers-and-other-algorithmic-problems.

284   humans are remarkably good at: J. N. MacGregor and T. Ormerod, "Human Performance on the Traveling Salesman Problem," *Perception & Psychophysics* 58, no. 4 (1996): 527–39.

INDEX

on container ships during pandemic, 28; Nazi slave laborers, 144; robots as slaves, 219; scientific management (Taylorism) as form of, 232; wage slavery of long-haul truckers, 157
slow steaming, 35–36
slungshot, 65
"smalls," 260
smartphones: delivery drivers, communication with, 284, 286; Moore's law and, 153; positional systems in, 145, 146; rapid delivery systems for, 265; Starship robotic delivery using computing power of, 268; supply-chain miles required to assemble, 12; truck drivers and, 120; Vietnam, manufacturing in, 16
Smokey and the Bandit (film, 1977), 109
Soft Robotics, 246, 247
sortation centers, 252–61
South China Sea, navigation in, 37–40
space flight, 144, 145, 247–48
stack ranking (rank and yank), 204
Stalin, Joseph, 61
"standard work," concept of, 230
Staples, 168, 182, 212
Star Wars (film, 1977), 109
Starbucks, 123, 263–65
Starship Technologies, 263–70
status competition, 91
steel production, 99–100
Stein, Gertrude, 98
Stone, Brad, *The Everything Store*, 170, 275
stopwatches, 96–97
stressful working conditions: at Amazon, 171–76, 191, 214–16, 234; automation, surveillance, and work intensification, 113, 157, 174–75, 203, 211–14, 231–32, 234–35; scientific management and, 88, 95, 97, 98, 213, 234; shipping crews affected by Covid-19 pandemic, 27–29; for truck drivers, 110–13, 117, 120–21, 125, 135, 157; turnover and, 113

strikes and striking, 33, 71, 88, 98, 277
STS (ship-to-shore) cranes, 20, 23, 34, 68–71, 77–78, 81
Suez Canal, 38, 91
supply chain, 1–12; consumer culture and, 1; Covid-19 pandemic, effects of, 2, 6–9, 10; factory system, incorporation into, 2, 90–92; Southeast Asia's role in, 10–12; "supply shocks," 9–12. *See also* delivery of goods to consumers; management systems; ships and shipping; trucks and truck drivers; warehouses and warehousing
surveillance: combined with work intensification, and automation, 113, 157, 174–75, 203, 211–14, 231–32, 234–35; of delivery drivers, 26; of workers for behavior leading to injury, 238, 286
surveillance capitalism, 231
survivalists, 7
swarm robotics, 78
Swift (trucking company), 107, 112

Target, 109
tarmac, invention of, 129
Taylor, Frederick Winslow, 87–90, 93–98, 103, 104, 105, 113, 213; *The Principles of Scientific Management* (1911), 95, 97, 98; *Shop Management* (1903), 95
Taylor, Robert, 216
Taylorism. *See* scientific management
Teamsters, 171, 277–78
teraflops, 154
Terminal Island, port of Los Angeles, 45, 48, 51
Tesla, 154, 155
textile manufacture: modern supply chain, 91–92; 19th-century factories, 91, 212
Theodore, Nik, 233
Thompson, Ben, 211
tilt-tray sorters, 166, 181, 223

ABOUT THE AUTHOR

CHRISTOPHER MIMS writes Keywords, a weekly technology column, for the *Wall Street Journal*. Before joining the *Journal* in 2014, he was the lead technology reporter for Quartz, and has written on science and tech for *MIT Technology Review*, *Smithsonian*, *Wired*, *The Atlantic*, and *Scientific American*, among other outlets. He has a degree in neuroscience and behavioral biology from Emory University and lives in Baltimore, Maryland.